STUDENT'S SOLUTIONS MANUAL

ELKA BLOCK

FRANK PURCELL

FUNCTIONING IN THE REAL WORLD: A PRECALCULUS EXPERIENCE

SECOND EDITION

Sheldon P. Gordon
Farmingdale State University of New York

Florence S. Gordon
New York Institute of Technology

Alan C. Tucker
SUNY at Stony Brook

Martha J. Siegel
Towson State University

PEARSON

Addison
Wesley

Boston   San Francisco   New York
London   Toronto   Sydney   Tokyo   Singapore   Madrid
Mexico City   Munich   Paris   Cape Town   Hong Kong   Montreal

Reproduced by Pearson Addison-Wesley from electronic files supplied by Twin Prime Editorial.

Copyright © 2004 Pearson Education, Inc.
Publishing as Pearson Addison-Wesley, 75 Arlington Street, Boston, MA 02116

ISBN   0-201-61137-6

Table of Contents

Chapter 1 Functions in the Real World

Section 1.1   Functions Are All Around Us

1.   **a.** Function, where the number of miles driven depends on the number of gallons used. This is a function because the number of gallons of gas gives an unique number of miles.

    b. Function, where the price depends on the number of carats. This is a function because the number of carats in a diamond determines exactly one price.

    c. Not a function, where the identity of a major league player depends on the number of home runs scored at the end of the season. This is not a function because two or more players can have the same number of home runs.

    d. Not a function, where the identity of a student depends on a specific SAT score in a particular year. This is not a function because two or more students can have the same SAT score.

    e. Function, where the amount of rain depends on the day. This is a function because each day has an unique amount of rainfall.

    f. Not a function, where the day of the year depends on the given amount of snowfall in Buffalo. This is not a function because two or more days can have the same amount of snowfall.

3.   In all three of the graphs we you could point out that clearly half-time and time-outs are not represented — these situations would produce sharp spikes and lulls on the noise levels. These are reasonable objections, and you could consider them to produce more realistic plots. Graph **i** matches Scenario **c**; Graph **ii** matches Scenario **a**; and Graph **iii** matches Scenario **d**. One of the graphs that Scenario **b** might produce is:

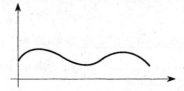

5.   To produce the graphs called for in Scenarios **a–d**, one must make assumptions about the path of travel. The simplest assumption is that travel is radially toward or away from home. The following graphs make that assumption, and assume home is in the city so that travel "towards town" is also towards home — suburbanites will need to adjust the graph of Scenario **b**. Also, sharp corners in the following graphs indicate more or less instantaneous changes of speed or direction. Make explicit any other assumptions you use in your own answers.

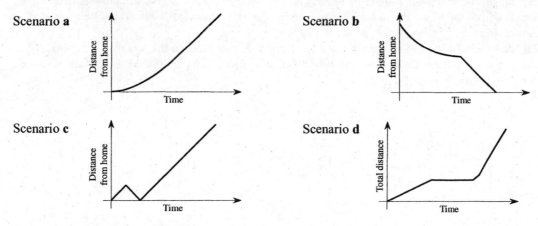

7.   Answers should vary considerably.

9. There are a number of levels at which an analysis can proceed. Clearly one would want to invest the year before the Dow-Jones has a peak and to divest oneself of stocks before a drop. As an aid to our analysis, we form a table of the differences in the value in a given year and the value in the subsequent year to see the rise or fall of the Dow-Jones average.

80	81	82	83	84	85	86	87	88	89
125	−89	172	212	−47	335	349	43	230	584

90	91	92	93	94	95	96	97	98	90
−119	535	132	457	76	1343	1270	1518	1219	2174

Looking at the changes in the Dow, one sees that 1985, 1986, 1989, 1991, 1993, 1995, 1997, and 1999 would have been good years to invest. More important than the changes in the Dow-Jones average, however, would be the percentage change. The table of percentage changes shown on the next page is:

80	81	82	83	84	85	86	87	88	89
15	−9	20	20	−4	28	23	2	12	27

90	91	92	93	94	95	96	97	98	90
−4	20	4	14	2	35	25	24	15	24

This table marks 1995, 1985, and 1989 as the three best years for investors.

Section 1.2   Describing the Behavior of Functions

1. a. The cost of postage either remains constant or increases at the beginning of each year, so it's neither strictly increasing nor strictly decreasing.
    b. The sunrise times can be earlier or later on different days, so the function is neither.
    c. The high temperatures can be higher or lower on different days, so the function is neither.
    d. The closing price of one share of IBM stock can be higher or lower on different days, so the function is neither.
    e. As the length of the base increases, the area of the triangle increases, so the function is strictly increasing.
    f. The height of the bungee jumper increases and decreases over time, so the function is neither.
    g. The height of the liquid in the tank keeps decreasing as the liquid leaks out, so the function is strictly decreasing.
    h. The daily cost of heating a home increases or decreases over time, so the function is neither.
    i. Each new record time for running the 100-meter dash is less than the previous record time, so the function is strictly decreasing.

3.
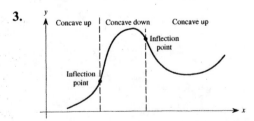

5.

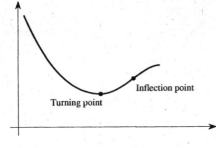

Turning point   Inflection point

7.

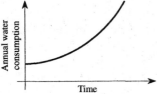

Annual water consumption / Time

9.

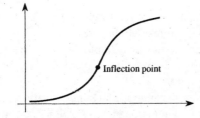

Inflection point

11. **a.** (Answers will vary.) The level of pollutants is increasing at a decreasing rate, so the company could say that they are responding to the need for less pollution by slowing down the increase in pollution. The EPA could counter this position by saying that the company is not responding because the pollution levels are still increasing.

b. The level of pollutants is increasing at an increasing rate. The EPA could say that the company is irresponsibly ignoring the need for less pollution and is endangering the environment.

c. The level of pollutants is decreasing at a decreasing rate. The company could say that they cannot reduce the level of pollutants as quickly as they would like, but they are still responding positively to the need for less pollution.

d. The level of pollutants is decreasing at an increasing rate. The company could say that not only are they responding to the need for less pollutions, they are doing so expeditiously.

13. **a.**

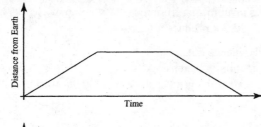

Distance from Earth / Time

b.

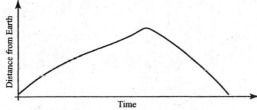

Distance from Earth / Time

15. Functions **a, b, c, d, e, h, i,** and **j** are periodic. Functions **f** and **g** are not periodic.

17.

19. **a.** The period of the sunspot cycle is roughly $10\frac{1}{2}$ years long.
   b. The next two peaks could be expected about 2000 and 2011.
   c. Using the average of the other observed peaks, we might estimate that the maximum number of sunspots at the next peak will be around 150.

Section 1.3   Representing Functions Symbolically

1. In several cases, whether the described relation is a function will depend on the interpretation of the intended range. A careful discussion of domains may also be necessary to resolve disputes.
   a. A function, provided we consider only one nation's postage as the dependent variable.
   b. Not a function if the range is taken as weight. If we consider the domain to be intervals of weight, the relation may be turned into a function.
   c. A function with the date as the independent variable, if we consider the sunrise at a single location.
   d. At a seaside location there are two high tides in any 24 hour period. Unless one tide is higher and thus the "true high tide" for a particular day, the relation is not a function.
   e. For the same location, the relation is a function with the date as the independent variable.
   f. A function, with the day as the independent variable and price as the dependent variable.
   g. Not a function. Two rectangles with the same base, but different heights, will have different areas.
   h. A function. Either area or height can be considered the independent variable.
   i, A function, although the form of the function will vary for differing interpretations of the dependent variable "height."
   j. Not a function. Since the whole point of a bungee jump is to bounce up and down, many heights will be repeated at different times.
   k. If we interpret the range to be *individual* ball players, the described relation is not a function. If we interpret the range to be *lists* of ball players the described relation is a function.
   l. A function. Either time or the height of liquid can be taken as the independent variable.
   m. The *average* cost of heating an *average* home could arguably be considered a function -- the word "average" can be interpreted in many ways, and level billing from the power company works on this premise. But an *actual* home with *actual* costs will not produce a function.

3. **a.** $f(10) = 50$ tells us that a 10-year old child is 50 inches tall.
   b. As a child ages, the child grows taller, so the function is increasing.
   c. The function is generally concave down because a child's growth starts quickly but then usually slows down.

5. $D(T) = kT$ for some constant k.

7. $F(d) = \dfrac{k}{d^2}$ for some constant k.

9. **a.** $f(1) = 156$ calories. If one gram of peanut butter is used, the sandwich will have 156 calories.
    b. $f(10) = 210$ calories, $f(15.5) = 243$ calories, $f(20) = 270$ calories, and $f(30) = 330$ calories.
    c. The number of calories from the bread alone is $f(0) = 150$ calories.
    d. A sandwich for which $P = -1$ would be one that itself consumed a gram of peanut butter, or at least the calories of one gram of peanut butter. A person cannot consume -1 grams of peanut butter.
    e. A reasonable domain might arguably be $0 \le P \le 50$. (50 grams of peanut butter is a lot!) The range corresponding to this domain would be $0 \le f(P) \le 450$.

11. **a.** A sandwich with 24 grams of peanut butter and 20 grams of jelly has $150 + 6(24) + 2.7(20) = 348$ calories.
    b. If we require that P and J are integers, there are two reasonable solutions. A sandwich with no jelly and 25 grams of peanut butter is one. A second, more flavorful sandwich is one with 20 grams of jelly and 16 grams of peanut butter.
    c. Since one gram of peanut butter provides 6 calories and one gram of jelly provides 2.7 calories, peanut butter is more caloric. The coefficients of P and J in the formula give the number of calories in each gram.

13. **a.**
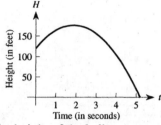

    b. The height of the ball at 1 second would be $H(1) = 120 + 60 - 16 = 164$ feet. The height of the ball at 4 seconds would be $H(4) = 120 + 240 - 256 = 104$ feet.
    c. The height of the ball two seconds after it has been thrown is given by $H(2) = 176$ feet. The height of the ball three seconds after it has been thrown is given by $H(3) = 156$ feet.
    d. The ball reaches its maximum height of 176.25 feet after 1.875 seconds.
    e. The ball hits the street roughly 5.19 seconds after it is thrown.
    f. The domain of H is given by $0 \le t \le 5.19$. The range of H is given by $0 \le H \le 176.25$.

15. $F(0) = 1/(0^2 - 4) = -1/4$, $F(1) = -1/3$, $F(3) = 1/5$, $F(4) = 1/12$, $F(5) = 1/21$. Evaluating the denominator of $F(x)$ at $x = 2$ would result in division by zero, which is undefined. The function F is defined (can be evaluated) for all other real numbers, so no other values of x should be skipped. The domain of F is the set of all real numbers except 2 and -2.

17. $g(4) = 4 + \sqrt{4} = 6$, $g(16) = 20$, $g(25) = 30$, $g(100) = 110$. Since we cannot take the square root of a negative number, the domain of $g(s)$ consists of all real numbers greater than or equal to zero (that is, $s \ge 0$). Since s is nonnegative and $\sqrt{s}$ is always nonnegative, the sum of s and $\sqrt{s}$ cannot be negative. So the range of g is a subset of the nonnegative real numbers.

19. $f(3) = 3^3 - 8(3^2) + 15(3) - 1 = -1$ $f(4) = 4^3 - 8(4^2) + 15(4) - 1 = -5$

    $f(5) = 5^3 - 8(5^2) + 15(5) - 1 = -1$ $f(3.5) = -3.625$

    $f(4.5) = -4.375$

    There are, of course, infinitely many values for which $f(x) < 0$.

Section 1.4   Connecting Geometric and Symbolic Representations

1. Estimates when reading graphs will vary.
   (i) The graph represents a function with domain $-3 \leq x \leq 4$ and a range $0 \leq y \leq 2$. It is decreasing from -3.5 to -2.5 and again from -1 to 1.7. It is increasing elsewhere. The graph is concave down from -1.8 to 0 and concave up from -3 to -1.8 and from 0 to 4.
   (ii) The graphs represents a function with domain $0 \leq x \leq 5$ and range $-1 \leq y \leq 2$. It is increasing and concave down everywhere.
   (iii) The graph does not represent a function.
   (iv) The graph does not represent a function.
   (v) The graph represents a function with domain $-1 \leq x \leq 5$ and range $-1 \leq y \leq 2$. It is decreasing and concave up everywhere.

3. **a.** The domain values t have been placed on the vertical axis and the range values $f(t)$ have been placed on the horizontal axis.
   b. The scale on the horizontal axis is not uniform. Each of the first three tick marks represent 10 units and the remaining tick marks represent 1 unit.
   c. The scale on the vertical axis is not uniform.
   d. The scale on the vertical axis is not strictly decreasing or strictly increasing. The scale goes from 0 to 400 and then decreases from 400 to 50.

5. For $f(x) = x^2 - 3x + 2$ and $x = -3, -2, -1, \ldots, 4, 5$ we obtain the following:

x	-3	-2	-1	0	1	2	3	4	5
f	20	12	6	2	0	0	2	6	12

$f\left(\frac{1}{2}\right) = \left(\frac{1}{2}\right)^2 - 3\left(\frac{1}{2}\right) + 2 = \frac{3}{4}$, $f\left(\frac{3}{2}\right) = -\frac{1}{4}$, $f\left(-\frac{5}{2}\right) = \frac{63}{4}$

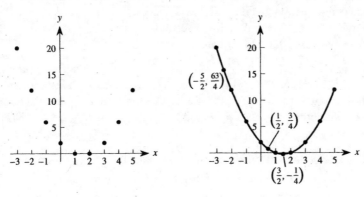

7. For $h(s) = s^3 - 7s + 5$, acceptable ranges need to encompass each of the three s-intercepts of h.

s	-4	-3	-2	-1	0	1	2	3	4
h	-31	-1	11	11	5	-1	-1	11	41

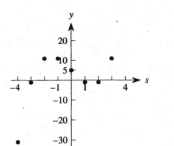 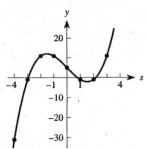

9. Since $f(x)$ values are increasing and concave up, $10 < f(5) < 20$ and $f(5) - f(4) < f(6) - f(5)$. If $f(5) = 15$, then $f(5) - f(4) = 15 - 10 = 5$ and $f(6) - f(5) = 20 - 15 = 5$. Since $f(5) - f(4) \neq f(6) - f(5)$, $f(5)$ must be less than 15. Thus, $f(5)$ can be any number between 10 and 15.

11. **a.** $x_1 < x < x_3$, $x_6 < x < x_8$, $x_{11} < x < x_{14}$; that is, (x_1, x_3), (x_6, x_8), (x_{11}, x_{14}).

  b. $x_4 < x < x_6$, $x_8 < x < x_{11}$; that is, (x_4, x_6), (x_8, x_{11})

  c. x_4, x_6, x_8, x_{11}

  d. x_4, x_8

  e. x_6, x_{11}

  f. x_3, x_5, x_7, x_9, x_{13}

  g. $x_1 < x < x_3$, $x_5 < x < x_7$, $x_9 < x < x_{13}$; that is, (x_1, x_3), (x_5, x_7), (x_9, x_{13})

  h. $x_3 < x < x_5$, $x_7 < x < x_9$, $x_{13} < x < x_{14}$; that is, (x_3, x_5), (x_7, x_9), (x_{13}, x_{14})

  i. $x_{12} < x < x_{14}$; that is, (x_{12}, x_{14})

  j. $x_9 < x < x_{10}$; that is, (x_9, x_{10})

  k. x_2, x_{10}, x_{12}

13. **a.** $g(x)$ because the functional values are increasing at a decreasing rate.

  b. $f(x)$ because the functional values are increasing at an increasing rate.

  c. $h(x)$ because the functional values are increasing at a constant rate.

15. **a.**

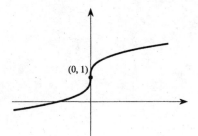

  b.

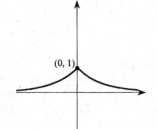

c.

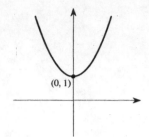

d.

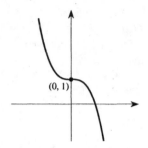

17. **a.** $f(x) = x^2 - 4$ is increasing for $x \geq 3$.

  b. $f(x) = x^2 - 4$ is always concave up.

19.

x	$f(x) = \sqrt{x}$	x	$f(x) = \sqrt{x}$
0	0	4	2
1	1	5	$\sqrt{5} \approx 2.236$
2	$\sqrt{2} \approx 1.414$	6	$\sqrt{6} \approx 2.449$
3	$\sqrt{3} \approx 1.732$		

$f\left(\dfrac{1}{2}\right) = \sqrt{\dfrac{1}{2}} \approx 0.707$

$f\left(\dfrac{3}{2}\right) = \sqrt{\dfrac{3}{2}} \approx 1.225$

$f\left(\dfrac{5}{2}\right) = \sqrt{\dfrac{5}{2}} \approx 1.581$

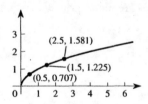

The domain of $f(x) = \sqrt{x}$ is $x \geq 0$.

Section 1.5   Mathematical Models

1. Since a chicken is a large object that does not conduct heat instantaneously, as the chicken cooks, different parts of the chicken will have different temperatures. Consequently, there is some question as to just what the "internal temperature of the chicken" might mean. However, cookbooks and the USDA recommend the internal temperature of $180°$, and so will we. The graph below assumes a three- to four- pound bird. The graph shows an increasing function that is increasing at a decreasing rate, and thus is concave down. A reasonable domain might be from $t = 0$ to $t = 80$ min, which is the time the internal temperature reaches $180°$. The range for T, the internal temperature, would be from $T = 70°$ to $T = 80°$. The graph shows a function that is increasing and concave down.

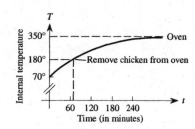

3. A typical graph would look like the one below showing a slight rise as the diver jumps up from the board, descends to water level and below, and then rises to the surface at a rate slower than the descent. The domain is roughly 0 to 4 seconds, and the range is roughly −3 to 11 meters.

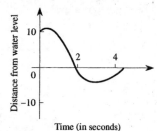

5. Using $s = f(d) = \sqrt{24d}$ to estimate the speed s from the length of a skid mark d, we get the following speeds:
   a. $f(60) \approx 38$ mph
   b. $f(100) \approx 49$ mph
   c. $f(140) \approx 58$ mph
   d. $f(200) \approx 69$ mph
   e. $f(d) = \sqrt{24d} \Rightarrow 60 = \sqrt{24d} = 3600 = 24d \Rightarrow 150$. So your skid marks will be approximately 150 ft long when making a panic stop from 60 mph.

7. a. $y = f(t) = 80 - 16t^2$
   b. $y = f(t) = v_0 t - 16t^2$

Chapter 1 Review Problems

1. Since the physician has no control over the depth of the tumor, it must be considered the independent variable. The amount of radiation needed depends on the depth of the tumor, so it is the dependent variable.

2. After the lead batter of the visiting team grounds out, the second batter hits a homer to put the visitors up by two. However, a relief pitcher is brought in and promptly strikes out the next two batters. The first hitter for the home team strikes out; however, the second hitter drops a stand up double into left field. The next batter lays down a sacrifice bunt to advance the runner to third. Another double hit scores the runner and puts the tying run on second. Alas, mighty Casey strikes out again retiring the side and ending the game.

3.

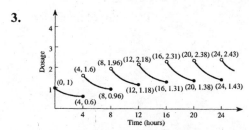

4. One such graph is the following. A key point will be the interpretation of the phrase "to grow steadily." Other interpretations are possible.

5. For both tables, each *x*-value has exactly one corresponding *y*-value. So both tables represent functions.

6. There is certainly a pattern of increasing attendance with increasing budget, as shown in the graph on the next page in which the horizontal axis represents budget and the vertical axis represents attendance.. Overall, zoos with larger budgets have greater attendance.

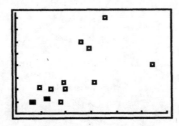

7. **a.**

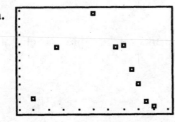

     Note:
The horizontal axis goes from 1980 to 2000 with scale 2.

The vertical axis goes from 150 to 450 with scale 25.

   b. The maximum number of cassettes occurred in approximately 1990.

   c. The most rapid change in the cassettes sold is between 1994 and 1995; the slowest change is between 1993 and 1994.

8. For $f(x) = 3x^2 - 2x + 1$, we have $f(0) = 0 - 0 + 1 = 1$, $f(1) = 3 - 2 + 1 = 2$,
$f(1.1) = 3(1.1)^2 - 2(1.1) + 1 = 2.43$, $f(1.01) = 3(1.01)^2 - 2(1.01) + 1 = 2.0403$,
$f(-3) = 27 + 6 + 1 = 34$, and $f(a) = 3a^2 - 2a + 1$.

9. **a.** $C = f(t) = 659.7t + 15,598$
     For 1995, $t = 5$: $C = f(5) = 659.7(5) + 15,598 = \$18,896.50$

   b. For 2003, $t = 13$: $C = f(13) = 659.7(13) + 15,598 = \$24,174.10$

   c. $20,000 = 659.7t + 15,598 \Rightarrow 659.7t = 4402 \Rightarrow t = 4402 / 659.7 \approx 6.7$. It will take approximately 6.7 years for the average cost of a new car to reach \$20,000.

10.

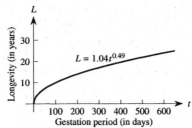

a. The model predicts a life span of $f(31) = 1.04(31)^{0.49} \approx 5.6$ years.

b. Since $f(15) \approx 3.9$ and $f(645) \approx 24.8$, the predicted range of longevity is from 3.9 to 24.8 years.

c. The graph is increasing and concave down.

d. We find the intersection point of the graph of $L = 1.04t^{0.49}$ and the graph of $L = 15$, which is at about $(231.97, 15)$. So the model predicts a gestation period of about 232 days for animals whose average longevity is 15 years.

e. The predicted longevity of humans is $f(270) = 1.04(270)^{0.49} \approx 16.2$ years. There are a variety of cultural and biological reasons that make this a low estimate.

11. **a.** The turning point is the vertex (minimum value) of the parabola $y = 4x^2 + 3x - 5$, which is approximately $(-0.375, -5.5625)$.

b. The range of $f(x) = 4x^2 + 3x - 5$ is $y \geq -5.5625$.

12. **a.** The point of inflection for $f(x) = x^3 - 9$ is $(0, 9)$.

b. The range of $f(x) = x^3 - 9$ is all real numbers.

13. **a.** The domain of $f(x) = \sqrt{x+5}$ is the set of real numbers x satisfying $x \geq -5$ since the square root function will not accept negative input.

b. Because the square root function will not accept negative input, x must satisfy the inequality $x^2 - 16 \geq 0$. So, the domain of $f(x) = \sqrt{x^2 - 16}$ is given by $x \leq -4$ or $x \geq 4$.

c. The domain of $g(x) = \dfrac{x^2 + 4}{x^2 - 9}$ is the set of all real numbers except 3 and -3 since evaluating g at these values would result in 0 in the denominator.

d. The domain of $g(x) = \dfrac{x^2 - 4}{x^2 + 9}$ is the set of all real numbers.

14. **a.**

Weight (ounces)	0–1	1–2	2–3	3–4	4–5
Postage (dollars)	0.37	0.60	0.83	1.06	1.29

b.

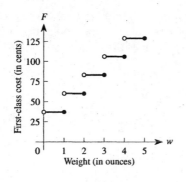

15. **a.** From A to B, the track is increasing and concave up.
    From B to C, the track is increasing and concave down.
    From C to D, the track is decreasing and concave down.
    From D to E, the track is decreasing and concave up.
    From E to F, the track is increasing and concave up.
    b. From A to B, the car's speed is decreasing at an increasing rate.
    From B to C, the car's speed is decreasing at an decreasing rate.
    From C to D, the car's speed is increasing at an increasing rate.
    From D to E, the car's speed is increasing at an decreasing rate.
    From E to F, the car's speed is decreasing at an increasing rate.

16. **a.**

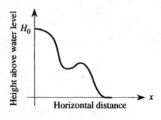

   b.

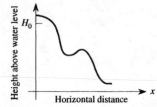

   c.

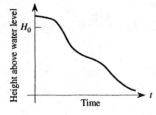

   d.

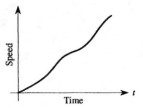

17. Answers will vary. The graph indicates an extremely high number every fourth year, so this suggests that the period is 4 years. Since the salmon die after spawning and there is a significant drop in the number of salmon the year following the peak, the graph suggests that the life span of the chinook salmon is 4 years.

Chapter 2   Families of Functions

Section 2.2   Linear Functions

1. **a.** The graph of $y = x + 2$ matches (iii) since both have a positive slope and a positive y-intercept.
    b. The graph of $y = x - 3$ matches (i) since both have a positive slope and a negative y-intercept.
    c. The graph of $y = -2x + 4$ matches (v) since both have a negative slope and a positive y-intercept.
    d. The graph of $y = -3x - 4$ matches (vi) since both have a negative slope and a positive y-intercept.
    e. The graph of $y = \frac{1}{2}x$ matches (iv) since both have a positive slope and y-intercept 0.
    g. The graph of $y = 3$ matches (ii) since both represent constant functions.

3. **a.** $m = \dfrac{5 - (-2)}{2 - 1} = 7$; $\ y - y_0 = m(x - x_0) \Rightarrow y - (-2) = 7(x - 1) \Rightarrow y + 2 = 7x - 7 \Rightarrow y = 7x - 9$

    b. $m = \dfrac{-2 - (-2)}{3 - 1} = 0$; $\ y - y_0 = m(x - x_0) \Rightarrow y - (-2) = 0(x - 1) \Rightarrow y + 2 = 0 \Rightarrow y = -2$

    c. $m = \dfrac{8.36 - 4.96}{-1.91 - 3.52} = \dfrac{3.4}{-5.43} \approx -0.62615$;
    $y - y_0 = m(x - x_0) \Rightarrow y - 4.96 = -0.62615(x - 3.52) \Rightarrow y - 4.96 = -0.62615x + 2.18768 \Rightarrow$
    $y = -0.62615x + 7.164$

5. **a.** Let $t = 0$ in 1990. Using $(0, 442.2)$ and $(8, 158.5)$, we get $m = \dfrac{158.5 - 442.2}{8 - 0} = -35.4625$;

    $CT - 442.2 = -35.4625(t - 0) \Rightarrow CT = -35.4625t + 442.2$

    b. Let $t = 0$ in 1990. Using $(0, 865.7)$ and $(8, 1124.3)$, we get $m = \dfrac{1124.3 - 865.7}{8 - 0} = 32.325$;
    $CD - 865.7 = 32.325(t - 0) \Rightarrow CD = 32.325t + 865.7$

    c. The sales of the cassette tapes are falling by 35.4625 million per year, and the sales of CDs are rising by 32.325 million per year.
    d. $CD > CT \Rightarrow 32.325t + 865.7 > -35.4625t + 442.2 \Rightarrow 67.7875t > -423.5 \Rightarrow t \approx -6.2$. The number of CDs sold overtook the number of cassette tapes sold in approximately 6 years and 2 months before 1990, that is sometime in late 1983.
    e. Using the combined data values, we now have $(0, 1307.9)$ and $(8, 1282.8)$, and so
    $m = \dfrac{1282.8 - 1307.9}{8 - 0} = \dfrac{-25.1}{8} = -3.1375$; letting $B = CT + CD$, we get
    $B - 1307.9 = -3.1375(t - 0) \Rightarrow B = -3.1375t + 1307.9$.
    f. If both the sales of cassette tapes and CDs were linear, then in 1995 ($t = 5$), we would have from our models that $CT = -35.4625(5) + 442.2 \approx 264.9$ and $CD = 32.325(5) + 865.7 \approx 1027.3$. Since the actual number of cassette tapes sold in 1995 is 272.6 million and the actual number of CDs sold in 1995 is also 272.6 million, the assumption that the sale trends were linear is incorrect.

7. **a.** $C = 0.30t + 0.40$
    b. The slope indicates that each minute cost $.30; the vertical intercept (initial cost at $t = 0$) indicates that it costs $.40 just to place the call.
    c. $C = 0.30(26) + 0.40 = \$8.20$
    d. The rate of $.30 discounted at 30% gives a new rate of $\$.30 - \$.30(.30) = \$.30 - \$.09 = \$.21$, and the cost to place a call of $.40 discounted at 30% gives a new cost of $\$.40 - \$.40(.30) = \$.40 - \$.12 = \$.28$;
    $C = 0.21t + 0.28$; each minute costs $.21, and it costs $.28 to place a call; $C = 0.21(26) + 0.28 = \$5.74$

9.   **a.** The cost function for DJ1 is $C = 60t + 120$, and the cost function for DJ2 is $C = 75t + 100$, where C represents the cost in dollars and t represents time in hours.

    b.

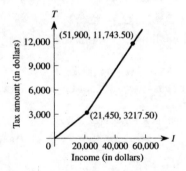

    c. Finding the intersection point $\left(\frac{4}{3}, 200\right)$ tells us that both DJs cost the same if the party lasts $\frac{4}{3}$ hr, or 1 hr 20 min. Looking at the graphs, we see that the graph for DJ1 is below the graphs of DJ2 when $t > \frac{4}{3}$. So, DJ1 will cost less than DJ2 if DJ1 is hired for longer than 1 hr 20 min.

11.   **a.** Taking T to represent the computed tax and I to represent the taxable income, we have the equation
$T = 3217.50 + 0.28(I - 21,450) = 0.28I - 2788.5$ for $21,450 \leq I \leq 51,00$. Since
$T = 0.28(21,450) - 2788.5 = 3217.50$ and $T = 0.28(51,900) - 2788.5 = 11,743.50$, the range for this function is $3217.50 \leq T \leq 11,743.50$.

    b. The slope is the percentage tax on income in the range, $21,450 \leq I \leq 51,900$. There is a 28% tax on income for single taxpayers with taxable income between \$21,450 and \$51,900.

    c. There is no gap between the two tax schemes although there is a fairly sharp increase in the rate of taxation (as was intended). See the graph below.

13.   **a.** Taking T to represent time after 1900 and N to represent the number of kilometers northward from the U.S.–Canadian border, the equation that describes the position of the Athabasca glacier is
$N = 15T + 300$.

    b. When $N = 0$, we get $0 = 15T + 300 \Rightarrow 15T = -300 \Rightarrow T = -20$; the toe of the glacier extended over the border 20 years before 1900, or about 1880.

    c. No, on the scale of a million years, one quickly runs out of both glacier and northward from the border.

15. **a.** A simplistic but unrealistic analysis would produce a graph like the one below on the left. The graph below on the right would be more realistic since the cyclist does not change speed instantly.

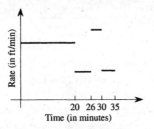
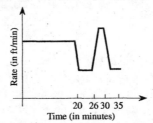

b. The total distance the bicyclist has traveled (considering only the intervals when the bicyclist's rate was constant) is $D = 1000 \times 20 + 500 \times 6 + 1200 \times 4 + 500 \times 5 = 30,300$ feet.

c.

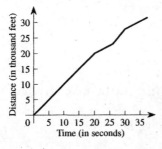

d. The formulas for distance D can be summarized as:

$$D = 1000t \qquad\qquad\qquad\qquad\qquad\qquad\quad \text{for } 0 \le t < 20$$
$$D = 20,000 + 500(t - 20) = 500t + 10,000 \qquad \text{for } 20 \le t < 26$$
$$D = 23,000 + 1200(t - 26) = 1200t - 8200 \qquad \text{for } 26 \le t < 30$$
$$D = 27,800 + 500(t - 30) = 500t + 12,800 \qquad \text{for } 30 \le t \le 35$$

17. **a.** Parallel lines have equal slopes, so the line parallel to $y = 5x - 3$ has slope 5. The equation of the parallel line passing through (6, 4) is $y - 4 = 5(x - 6) \Rightarrow y - 4 = 5x - 30 \Rightarrow y = 5x - 26$.

b. The slopes of perpendicular lines are negative reciprocals, so any line perpendicular to $y = 5x - 3$ has slope $-1/5$. The equation of the perpendicular line passing through (6, 4) is $y - 4 = -\frac{1}{5}(x - 6) \Rightarrow y = -\frac{1}{5}x + \frac{26}{5}$.

19. **a.**

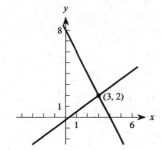

b. Solve the equations simultaneously using the elimination (or addition) method:

$$\left.\begin{array}{l} 3x-4y=1 \\ 2x+\ y=8 \end{array}\right\} \xrightarrow[\text{multiply 2nd eq. by 4}]{} \begin{array}{l} 3x-4y=1 \\ \underline{8x+4y=32} \\ 11x\quad\ =33 \\ \quad\quad x=3 \end{array}$$

$x=3 \Rightarrow 2x+y=8 \Rightarrow 2(3)+y=8 \Rightarrow y=2;$ the point of intersection is $(3, 2)$.

c. Plotting the line $7x-2y=17$, we see that all three lines intersect at $(3, 2)$.

d. Plotting the line $9x-y=25$, we see that all four lines intersect at $(3, 2)$.

e. The resulting line is the vertical line $x=3$. Eliminating the y-term gives a vertical lines that has the same x-value as the x-coordinate of the point of intersection, which is the x-value of the solution to the system.

f. Multiplying Equation (2) by $-\frac{3}{2}$ will eliminate the x-term. The resulting line is the graph of the horizontal line $y=2$, which is the y-value of the solution to the system.

21. a. (i) $3x-4y=12 \Rightarrow y=\frac{3}{4}x-3 \Rightarrow m=\frac{3}{4}$

         (ii) $2x+5y=10 \Rightarrow y=-\frac{2}{5}x+2 \Rightarrow m=-\frac{2}{5}$

         (iii) $8x+6y=7 \Rightarrow y=-\frac{4}{3}x+\frac{7}{6} \Rightarrow m=-\frac{4}{3}$

         Since the slope of **(i)** is the negative reciprocal of the slope of **(iii)**, these two function represent perpendicular lines.

     b. Solve the equations simultaneously using the elimination (or addition) method:

$$\left.\begin{array}{l} 3x-4y=12 \\ 8x+6y=7 \end{array}\right\} \xrightarrow[\text{multiply 2nd eq. by 4}]{\text{multiply 1st eq. by 6}} \begin{array}{l} 18x-24y=72 \\ \underline{32x+24y=28} \\ 50x\quad\ =100 \\ \quad\quad x=2 \end{array} \qquad \begin{array}{l} 3(2)-4y=12 \\ 6-4y=12 \\ \quad\ y=-1.5 \end{array}$$

The point of intersection is $(2, -1.5)$.

Exercising Your Algebra Skills

1. $5x-7=12 \Rightarrow 5x=19 \Rightarrow x=19/5$

2. $3x+8=-7 \Rightarrow 3x=-15 \Rightarrow x=-5$

3. $4x-3=-5 \Rightarrow 4x=-2 \Rightarrow x=-1/2$

4. $8x+7=15 \Rightarrow 8x=8 \Rightarrow x=1$

5. $18y-7=22 \Rightarrow 18y=29 \Rightarrow y=29/18$

6. $5.4x-7.2=0.8 \Rightarrow 5.4x=8.0 \Rightarrow x\approx1.48$

7. $9-3x=6 \Rightarrow -3x=-3 \Rightarrow x=1$

8. $5-4p=-7 \Rightarrow -4p=-12 \Rightarrow p=3$

9. $4.7q+5.1=24.5 \Rightarrow 4.7q=19.4 \Rightarrow q\approx4.1$

10. $-1.3w+12.8=22.7 \Rightarrow -1.3w=9.9 \Rightarrow w=-7.6$

11. $4k+7=9k-8 \Rightarrow 5k=15 \Rightarrow k=3$

12. $6z-5=4z+11 \Rightarrow 2z=16 \Rightarrow z=8$

13. $3(2x-5)=4 \Rightarrow 6x-15=4 \Rightarrow 6x=19 \Rightarrow x=19/6$

14. $2(4-3w)=7 \Rightarrow 8-6w=7 \Rightarrow -6w=-1 \Rightarrow w=1/6$

15. $3.2(t-1980)=1700 \Rightarrow 3.2t-6336=1700 \Rightarrow 3.2t=8036 \Rightarrow t=2511.25$

16. $1.35(t-75)=8 \Rightarrow 1.35t-101.25=8 \Rightarrow 1.35t=109.25 \Rightarrow t \approx 80.9$

17. **a.** $2x-3y=8 \Rightarrow y=\frac{2}{3}x-\frac{8}{3}$; $m=\frac{2}{3}$
    b. x-intercept: $2x-3(0)=8 \Rightarrow x=4$
    y-intercept: $2(0)-3y=8 \Rightarrow y=-\frac{8}{3}$

18. **a.** $2x+3y=8 \Rightarrow y=-\frac{2}{3}x+\frac{8}{3}$; $m=-\frac{2}{3}$
    b. x-intercept: $2x+3(0)=8 \Rightarrow x=4$
    y-intercept: $2(0)+3y=8 \Rightarrow y=\frac{8}{3}$

19. **a.** $4x+7y+5=0 \Rightarrow y=-\frac{4}{7}x-\frac{5}{7}$; $m=-\frac{4}{7}$
    b. x-intercept: $4x+7(0)=-5 \Rightarrow x=-\frac{5}{4}$
    y-intercept: $4(0)+7y=-5 \Rightarrow y=-\frac{5}{7}$

20. **a.** $3y-2x+4=0 \Rightarrow y=\frac{2}{3}x-\frac{4}{3}$; $m=\frac{2}{3}$
    b. x-intercept: $3(0)-2x=-4 \Rightarrow x=2$
    y-intercept: $3y-2(0)=-4 \Rightarrow y=-\frac{4}{3}$

Section 2.3   Linear Functions and Data

1. **a.** $\Delta y=-6,-6,-6,-6$, respectively. Since the difference between the functional values is constant, the function is linear. Since $\Delta y=-6$ and $\Delta x=1$, $m=-6/1=-6$. Using the point $(5,77)$, the equation of the line is $y-77=-6(x-5) \Rightarrow y=-6x+107$; for $x=10$, $y=-6(10)+107=47$.
   b. $\Delta y=0.7,0.7,0.6,0.7$, respectively. Since the difference between the functional values is not constant, the function is not linear.
   c. $\Delta y=2.4,2.4,2.4,2.4$, respectively. Since the difference between the functional values is constant, the function is linear. Since $\Delta y=2.4$ and $\Delta x=5$, $m=2.4/5=0.48$. Using the point $(75. 125.1)$, the equation of the line is $y-125.1=0.48(x-75) \Rightarrow y=0.48x+89.1$; for $x=100$, $y=0.48(100)+89.1=137.1$.

3. Since the tabular x values in the data table are uniformly spaced one unit apart, the slope is 0.057 (the difference of the consecutive y values). The point $(4, 1.557)$ is on the line, therefore an equation for the line is $y-1.557=0.057(x-4) \Rightarrow y=0.057x+1.329$.

5. **a.** $(0, 26.5)$ and $(10, 33.1) \Rightarrow m=(33.1-26.5)/10=0.66$; $y-26.5=0.66(x-0) \Rightarrow y=0.66x+26.5$
   b. In 2002, $t=22$, so $y=0.66(22)+26.5=41.02$ billion.
   c. If $t=0$ in 1900, then our new data points are $(80, 26.5)$ and $(90, 33.1)$; $m=(33.1-26.5)/(90-80)=0.66$ and $y-26.5=0.66(x-80) \Rightarrow y=0.66x-26.3$; in 2002, $t=102$, so $y=0.66(102)-26.3=41.02$ billion.

7. a.

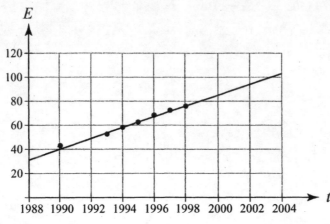

b. Using the points (1990, 40) and (1998, 76), we estimate the slope to be 4.5, which tells us that the total value of electronics and electronic components increased $4.5 billion dollars each year.

c. Our best estimate for the equation of the line is $E - 40 = 4.5(t - 1990) \Rightarrow E = 4.5t - 8915$.

d. The total value of electronic and electronic components in 2004 is $E = 4.5(14) - 8915 = \$103$ billion.

9. a.

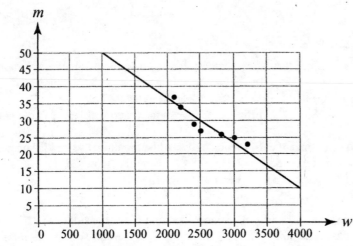

b. We get a reasonably good fit by using two of the actual data points, (2200, 34) and (2800, 26). We estimate the slope to be $-1/75$, which gives the equation $M - 26 = -\frac{1}{75}(W - 2800) \Rightarrow M = -\frac{1}{75}W + \frac{190}{3}$.

c. Using this equation, we get 32 mpg for 2350 lbs, 22 mpg for 3100 lbs, and 37.33 mpg for 1950 lbs.

d. Solving $32 = -\frac{1}{75}W + \frac{190}{3}$ for W, we already know from part (c) that the answer will be 2350 lbs.

11. a. If N is the number of pounds of nuts and G the number of pounds of Gummi Bear candy that will be bought exhausting the budget, then $3N + 2G = 30$. This equation is called the budget constraint.

  b. See graphs on next page.

  c. The practical domain is $0 \le N \le 10$. The corresponding range is $0 \le G \le 15$.

  d. The new budget constraint is $3N + 2G = 60$. See graphs on next page.

  e. Budget $= \$30$, price of Gummi Bears $= \$1$ per pound; budget constraint is $3N + G = 30$. See graphs on next page.

f. Budget = \$30, price of nuts = \$6 per pound; budget constraint is $6N + 2G = 30$. See graphs below.

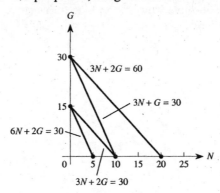

13. a. $3y - 2x = 12 \Rightarrow 3y = 2x + 12 \Rightarrow y = \frac{2}{3}x + 4$, the slope is $\frac{2}{3}$ and the vertical intercept is 4;

$4x + 5y = 20 \Rightarrow 5y = -4x + 20 \Rightarrow y = -\frac{4}{5}x + 4$, the slope is $-\frac{4}{5}$ and the vertical intercept is 4.

b.

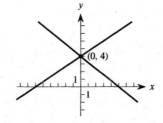

c. (i) See graph in part (b). **(ii)** Keep substituting different choices for x and y until you find the x and y values that satisfy both equations.

(iii) $\begin{aligned} -2x + 3y &= 12 \\ 4x + 5y &= 20 \end{aligned}\Bigg\}$ $\xrightarrow{\text{multiply 1st eq. by 2}}$ $\begin{aligned} -4x + 6y &= 24 \\ 4x + 5y &= 20 \\ \hline 11y &= 44 \\ y &= 4 \end{aligned}$ $\begin{aligned} -2x + 3(4) &= 12 \\ -2x &= 0 \\ x &= 0 \end{aligned}$

The solution is (0, 4).

15. The line L that passes through (70, 300) and (60, 250) has equation $y = 5x - 50$ and can be used as a reference to determine which functional values are possible for f.

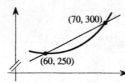

Any curve drawn through (60, 250) and (70, 300) that is concave up must lie below L for $60 < x < 70$ and above L for $x < 60$ and $x > 70$. When $x = 65$, we get a height of 275 along the line, so any possible values for $f(65)$ must be smaller than 275. The only appropriate choice is 270. When $x = 100$, we get a height of 450 along the line. So only values greater than 450 are possible for the function and, among the given choices, only 500 is possible. When $x = 40$, we get a value of 150 along the line. Since the curve is above the line there, the only possible value is 200. Thus, the possible values occur in **a, f,** and **i**; the impossible values occur in **b, c, d, e, g,** and **h**.

17.   **a.** $m_{\overline{PQ}} < m_{\overline{PR}} < m_{\overline{QR}} < 0$

      b. Perceived "steepness" is the absolute value of slope, so for negative slopes, "steepness" reverses the order of the slope values: $\text{steepness}_{\overline{QR}} < \text{steepness}_{\overline{PR}} < \text{steepness}_{\overline{PQ}}$

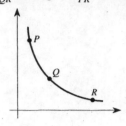

Exercising Your Algebra Skills

1. $A = bh \Rightarrow h = \dfrac{A}{b}$

2. $C = 2\pi r \Rightarrow r = \dfrac{C}{2\pi}$

3. $A = \pi r^2 \Rightarrow r^2 = \dfrac{A}{\pi} \Rightarrow r = \sqrt{\dfrac{A}{\pi}}$

4. $K = \frac{1}{2}mv^2 \Rightarrow 2K = mv^2 \Rightarrow m = \dfrac{2K}{v^2}$

5. $K = \frac{1}{2}mv^2 \Rightarrow 2K = mv^2 \Rightarrow v^2 = \dfrac{2K}{m} \Rightarrow v = \sqrt{\dfrac{2K}{m}}$

6. $F = \dfrac{GmM}{d^2} \Rightarrow d^2 = \dfrac{GmM}{F} \Rightarrow d = \sqrt{\dfrac{GmM}{F}}$

7. $T = 2\pi\sqrt{\dfrac{l}{g}} \Rightarrow \sqrt{\dfrac{l}{g}} = \dfrac{T}{2\pi} \Rightarrow \dfrac{l}{g} = \left(\dfrac{T}{2\pi}\right)^2 \Rightarrow l = g \cdot \left(\dfrac{T}{2\pi}\right)^2$

8. $F = \dfrac{mv^2}{r} \Rightarrow r = \dfrac{mv^2}{F}$

9. $F = \dfrac{mv^2}{r} \Rightarrow mv^2 = rF \Rightarrow v^2 = \dfrac{rF}{m} \Rightarrow v = \sqrt{\dfrac{rF}{m}}$

10. $4x - 5y = 20 \Rightarrow -5y = -4x + 20 \Rightarrow y = \frac{4}{5}x - 4; \; m = \frac{4}{5}$

11. $6x + 5y = 30 \Rightarrow 5y = -6x + 30 \Rightarrow y = -\frac{6}{5}x + 6; \; m = -\frac{6}{5}$

12. $5x - 4y = 10 \Rightarrow -4y = -5x + 10 \Rightarrow y = \frac{5}{4}x - \frac{5}{2}; \; m = \frac{5}{4}$

13. $2x + 7y = 9 \Rightarrow 7y = -2x + 9 \Rightarrow y = -\frac{2}{7}x + \frac{9}{7}; \; m = -\frac{2}{7}$

Section 2.4   Exponential Growth Functions

1.   **a.** Nation A has the greatest growth rate.

      b. Nation C has the smallest growth rate.

      c. Nations B and C have the largest initial population.

      d. Nation D has the smallest initial population.

      e. Nations B and D have the same growth rate; their growth curves are "parallel" to each other.

3.   A represents Anne's balance: $1200(1.04)^t$; B represents Bill's balance: $1000(1.045)^t$; C represents Christine's balance: $1500(1.038)^t$; D represents Doug's balance: $1200(1.045)^t$; E represents Elka's balance: $1300(1.0425)^t$. A simultaneous plot of the graphs of these balance functions is shown on the next page. Note that for clarity in the picture, the scale on the vertical axis does not begin at 0.

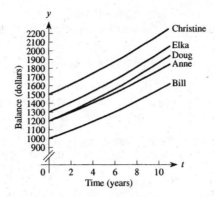

5. a. In 1990 we have $t = 0$, $(0, 495)$ and $(8, 912)$; for $(0, 495)$, $k = 495$ and $f(t) = 495c^t$;

for $(8, 912)$, $f(8) = 495c^8 = 912 \Rightarrow c^8 = 912/495 \approx 1.8424 \Rightarrow c = \sqrt[8]{1.8424} \Rightarrow c \approx 1.0794$;

$f(t) = 495(1.0794)^t$

b. In 1990 we have $t = 90$, $(90, 495)$ and $(98, 912)$; for $(90, 495)$, $f(t) = kc^t \Rightarrow f(90) = kc^{90} = 495 \Rightarrow$

$k = 495/c^{90}$; for $(98, 912)$, $f(98) = kc^{98} = 912 \Rightarrow \left(\dfrac{495}{c^{90}} \right) c^{98} = 912 \Rightarrow 495c^8 = 912 \Rightarrow$

$c = \sqrt[8]{\dfrac{912}{495}} \approx 1.0794$; thus, $k = 495/(1.0794)^{90} \approx 0.511$; $f(t) = 0.511(1.0794)^t$

c. In 1990 we have $t = 1990$, $(1990, 495)$ and $(1998, 912)$; for $(1990, 495)$, $f(t) = kc^t \Rightarrow$

$f(1990) = kc^{1990} = 495 \Rightarrow k = 495/c^{1990}$; for $(98, 912)$, $f(1998) = kc^{1998} = 912 \Rightarrow$

$\left(\dfrac{495}{c^{1990}} \right) c^{1998} = 912 \Rightarrow 495c^8 = 912 \Rightarrow c = \sqrt[8]{\dfrac{912}{495}} \approx 1.0794$; thus $k = 495/(1.0794)^{1990} \approx 4.59 \times 10^{-64}$;

$f(t) = (4.59 \times 10^{-64})(1.0794)^t$

d. The parameter k changed and the parameter c remained the same. Changing the starting value for the independent value only affects the initial value; it doesn't affect the growth factor.

e. In 2005, $t = 15$, $f(t) = 495(1.0794)^t \Rightarrow f(15) = 495(1.0794)^{15} \approx 1557.19$ billion

In 2005, $t = 105$, $f(t) = 0.511(1.0794)^t \Rightarrow f(105) = 0.511(1.0794)^{105} \approx 1558.02$ billion

In 2005, $t = 2005$, $f(t) = (4.59 \times 10^{-64})(1.0794)^t \Rightarrow f(2005) = (4.59 \times 10^{-64})(1.0794)^{2005} \approx 1557.95$ billio

7. a. In 1980, we have $t = 0$, and the data points $(0, 119)$ and $(14, 274)$; for $(0, 119)$,

$k = 119$ and $f(t) = 119c^t$;

for $(14, 274)$, $f(14) = 119c^{14} = 274 \Rightarrow c^{14} = 274/119 \approx 2.3025 \Rightarrow c = \sqrt[14]{2.3025} \Rightarrow c \approx 1.0614$;

so, $f(t) = 119(1.0614)^t$

b. In 1990, $t = 10$, so $f(10) = 119(1.0614)^{10} \approx \$215.94$ trillion.

c. In 2004, $t = 24$, so $f(24) = 119(1.0614)^{24} \approx \$497.33$ trillion.

d. $f(t) = \$500$ trillion $\Rightarrow 500 = 119(1.0614)^t \Rightarrow (1.0614)^t = 500/119$; using a function grapher with $y_1 = 1.0614^t$ and $y_2 = 500/119$, we get $t \approx 24.1$. So the total sales will reach $500 trillion in 2004.

9. a. For a growth rate of 2.6%, the growth factor is 1.026; the expression for the population P at any time t is $P = P_0(1.026)^t$. If t is the number of years since 1995, then we have $(0, 21.8)$ as a data point, and so $P = 21.8(1.026)^t$, where $t = 0$ in 1995.

b. In 2005, $t = 10$ and $P = 21.8(1.026)^{10} \approx 28.2$ million.

c. $21.8(1.026)^t = 2 \cdot 21.8 \Rightarrow (1.026)^t = 2 \Rightarrow t \approx 27$. The population of Venezuela will double about every 27 years.

11. a. Let t be the number of years since 1990. Since we have $(0, 1.36)$, $k = 1.36$ and $f(t) = 1.36c^t$; using the data point $(8, 1.595)$, we get $f(8) = 1.36c^8 = 1.595 \Rightarrow c = \sqrt[8]{1.595/1.36} \approx 1.0201$; so the function is $f(t) = 1.36(1.0201)^t$, where $t = 0$ in 1990.

b. In 2004, $t = 14$, $f(14) = 1.36(1.0201)^{14} \approx 1.80$ billion metric tons.

13. **a.** Let t be the number of years since 1999. A growth rate of 1.5% gives the growth factor 1.015. So the equation of the function representing the world's population is $P(t) = 6(1.015)^t$;
$f(15) = 6(1.015)^{15} \approx 7.5$ billion.

   b. $6(1.015)^t = 2 \cdot 6 \Rightarrow (1.015)^t = 2 \Rightarrow t \approx 46.6$. The world's population will double in about 47 years.

15. At 6% annual interest compounded quarterly, the quarterly rate is $1/4(6\%) = 1.5\%$. In t years, there are $4t$ quarters, therefore the balance after t years in the bank account if $100 is deposited is $100(1.015)^{4t}$. For $t = 1$ year, the balance is $100(1.015)^4 \approx \$106.15$; for $t = 10$ years, the balance is $100(1.015)^{40} \approx \$181.40$.

17. If the $450,000 had been loaned with the understanding that it would accrue 6% annual interest, then the amount owed in 1991 after 215 years would be $450,000(1.06)^{215} = \$124,157,667,678.78$, an amount well over 124 billion dollars.

19. **a.** The growth factor is computed as $a = f(8)/f(7) = 28.8/25.6 = 1.125$.

   b. The growth rate is computed as growth rate = growth factor − 1 = 0.125, or 12.5%.

   c. $f(10) = 1.125 \cdot f(9) = 1.125 \cdot 1.125 \cdot f(8) = 1.125^2 \cdot 28.8 = 36.45$

   d. $f(x) = k(1.125)^x, f(8) = k(1.125)^8 \Rightarrow 28.8 = k(1.125)^8 \Rightarrow k \approx 11.2; f(x) = 11.2(1.125)^x$

21. Letting $t = 0$ in 1981 gives the data points $(0, 964)$ and $(19, 11,358)$; $k = 964$ and $11,358 = 964c^{19} \Rightarrow$
$c = \sqrt[19]{11,358/964} \approx 1.1386; D = 964(1.1386)^t$; at the beginning of 2004, $D = 964(1.1386)^{23} \approx 19,081.57$.

23. The line containing the points $(0, 512)$ and $(4, 1250)$ represents the function $g(x) = 184.5x + 512$. An exponential function going through the same points $(0, 512)$ and $(4, 1250)$ is concave up, and so its graph must be below the line between $x = 0$ and $x = 4$. Since $g(2) = 184.5(2) + 512 = 881$ and $f(2) < g(2)$, we know that $f(2) = 800$ is possible and $f(2) \geq 881$ is impossible.

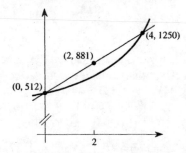

25. As in Problem 24, Acme's income is given by $I = 240(1.10)^t$. Now Finest's income is specified by $I = 300 + 25t$. Solving $240(1.10)^t = 300 + 25t$ for t by trial and error, we get an approximate value of 7.4 years. Thus Acme's income exceeds Finest's in the middle of 1997.

27. If the growth is exponential, the total growth of 3000% is the result of *compounding* an annual growth rate for 8 years. An annual growth rate of 375% for 8 years would produce a total growth much greater than 3000%. As Example 7 shows, the annual growth is actually about 152%.

Exercising Your Algebra Skills

1. $x^5 \cdot x^3 = x^{5+3} = x^8$

2. $x^4 \cdot x^2 = x^{4+2} = x^6$

3. $a^8 \cdot a^4 = a^{8+4} = a^{12}$

4. $\dfrac{a^{15}}{a^6} = a^{15-6} = a^9$

5. $x^{-5} \cdot x^3 = x^{-5+3} = x^{-2}$

6. $a^5 \cdot a^{-3} = a^{5-3} = a^2$

7. $\dfrac{r^8}{r^{-4}} = r^{8-(-4)} = r^{12}$

8. $\dfrac{b^{15}}{b^{-6}} = b^{15-(-6)} = b^{21}$

9. $\dfrac{w^{-4}}{w^{-7}} = w^{-4-(-7)} = w^3$

10. $\dfrac{w^{-7}}{w^{-4}} = w^{-7-(-4)} = w^{-3}$

11. $x^{-1/2}x^{3/4} = x^{(-1/2)+(3/4)} = x^{1/4}$

12. $y^{2/3}y^{4/3} = y^{(2/3)+(4/3)} = y^{6/3} = y^2$

13. $z^{2/3} z^{-5/3} = z^{-3/3} = z^{-1}$ **14.** $(x^5)^3 = x^{5 \cdot 3} = x^{15}$ **15.** $(x^3)^5 = x^{3 \cdot 5} = x^{15}$

16. $(a^8)^{-4} = a^{8(-4)} = a^{-32}$ **17.** $(a^3 b^5)^4 = a^{12} b^{20}$

18. $(a^3 + b)^2 = (a^3 + b)(a^3 + b) = a^6 + 2a^3 b + b^2$ **19.** $(a^3 - b)^2 = (a^3 - b)(a^3 - b) = a^6 - 2a^3 b + b^2$

Section 2.5   Exponential Decay Functions

1. Graphs **a**, **c**, and **f**, are exponential because they are concave up with the *x*-axis as the horizontal asymptote (that is, they eventually approach 0 asymptotically.) Graphs **b** and **e** are not exponential because they are concave down and have the functional value 0. Graph **d** is not exponential because it is concave down and has both positive and negative functional values.

3. An exponential function has values that are always positive or always negative. Since the points in part (b) are on opposite sides of the horizontal axis, they do not determine an exponential function. The ordered pairs in parts (a), (c) and (d) determine exponential functions. Following are sketches of the general shape that the pair of points would determine in each of these cases, as well as the information on the parameters *c* and *A* that they imply.

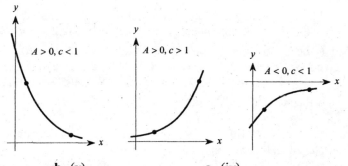

5. **a.** (vi)   **b.** (v)   **c.** (iv)   **d.** (ii)

7. **a.** Letting $t = 0$ in 1980 gives the data points (0, 27,700) and (17, 19,900); the point (0, 27,700) tells us that $k = 27,700$ and $y = kc^t \Rightarrow y = 27,700c^t$; the point (17, 19,900) tells us that
$$y = 27,700c^t \Rightarrow 19,900 = 27,700c^{17} \Rightarrow c = \sqrt[17]{19,900/27,700} \approx 0.9807; \text{ thus, } y = 27,700(0.9807)^t.$$
   b. Since $t = 24$ in 2004, $y = 27,700(0.9807)^{24} \approx 17,352$ cases.
   c. $y = 27,700(0.9807)^t \Rightarrow 10,000 = 27,700(0.9807)^t \Rightarrow (0.9807)^t = 10,000/27,700$; we solve for t graphically using $y_1 = (0.9807)^t$ and $y_2 = 10,000/27,700$: $t \approx 52.3$. So, the number of reported cases will drop to 10,000 in about 52.3 years.

9. **a.** If 35% of the nicotine is washed out, then $1 - 0.35 = 0.65$ of the nicotine remains. So the level of the nicotine in the blood has 0.65 as the decay factor. Since the initial value of nicotine is 0.4 mg, the equation of the function is $y = kc^t \Rightarrow y = 0.4(0.65)^t$.
   b. $y = 0.4(0.65)^t \Rightarrow 0.005 = 0.4(0.65)^t \Rightarrow (0.65)^t = 0.005/0.4 = 0.0125$; solving graphically for t gives $t \approx 10.2$. Thus, the amount of nicotine in the blood drops to 0.005 mg after 10.2 hours.

11. **a.** If $31\% = 0.31$ of the morphine is washed out, then $1 - 0.31 = 0.69$ of the morphine remains, and so 0.69 is the decay factor; since the initial amount given is 3 mg, the equation that models the function is
$$y = kc^t \Rightarrow y = 3(0.69)^t.$$
   b. $y = 3(0.69)^4 \approx 0.68$ mg.
   c. $0.2 = 3(0.69)^t \Rightarrow (0.69)^t = 0.2/3$; solving graphically for t gives $t \approx 7.3$. Thus, the amount of morphine left drops to 0.2 mg after about 7.3 hours.

13. Strontium 90 has a half-life of 29 years. In order to determine how long it takes for a sample contamination at 10 times the healthy level to decay to the health limit, we need to find t such that $(1 - 0.5)^t = 1/10$, that is, $t = 3.32$ half-lives, or 96.3 years.

15. a. If the filter removes 30% of the impurities, the decay rate is 0.30, and so the decay factor is 0.70; the equation is $I(t) = I_0(0.7)^t$, where $I_0 =$ the original amount of impurities.

　　b. $I(5) = I_0(0.7)^5 = 0.16807 I_0$. After 5 hours, about 16.8% of the impurities is left.

17. Let $R_0 =$ the initial number of kidney transplant recipients. For half-life $= 9.1$ years, we get
$$R(t) = R_0 c^t \Rightarrow \tfrac{1}{2}R_0 = R_0 c^{9.1} \Rightarrow c^{9.1} = \tfrac{1}{2} \Rightarrow c \approx 0.9268; \text{ for half-life} = 13.3 \text{ years, we get}$$
$$R(t) = R_0 c^t \Rightarrow \tfrac{1}{2}R_0 = R_0 c^{13.3} \Rightarrow c^{13.3} = \tfrac{1}{2} \Rightarrow c \approx 0.9524. \text{ Since decay rate} = 1 - \text{decay factor, for}$$
half-life $= 9.1$ years, we get decay rate $= 1 - 0.9268 = 0.0732 \approx 7.3\%$, and for half-life $= 13.3$ years, we get decay rate $= 1 - 0.9524 = 0.0476 \approx 4.8\%$. So, the later result from the 1996 study is good news since the decay rate is lower.

19. a. Let $S_0 =$ the initial number of patients who survive. $S(t) = S_0 c^t \Rightarrow S(5) = S_0 c^5 \Rightarrow 0.80 S_0 = S_0 c^5 \Rightarrow$
　　　$c^5 = 0.8 \Rightarrow c \approx 0.9563$. So the exponential function is $S(t) = S_0(0.9563)^t$;
　　　$S(10) = S_0(0.9563)^{10} = 0.6396 S_0$. The percentage of patients that will survive 10 years is about 64%.

　　b. $S = 0.956^t$

　　c. To find the half-life for survival, we solve $0.956^t = \tfrac{1}{2}$ for t, which gives $t \approx 15.4$. The half-life for survival is about 15.4 years.

21. a. Let $t = 0$ in 1990. The data point $(0, 442.2)$ tells us that $k = 442.2$; using the data point $(8, 158.5)$, we get $y = kc^t \Rightarrow y = 442.2c^t \Rightarrow 158.5 = 442.2c^8 \Rightarrow c = \sqrt[8]{158.5/442.2} \approx 0.8796$. So, the exponential function that models the number of cassette tapes sold is $T = 442.2(0.8796)^t$.

　　b. The data points given for CDs are $(0, 865.7)$ and $(8, 1124.3)$; thus, we have
　　　$y = 865.7c^t$ and $y = kc^t \Rightarrow 1124.3 = 865.7c^8 \Rightarrow c = \sqrt[8]{1124.3/865.7} \approx 1.0332$. So the exponential function that models the number of CDs sold is $C = 865.7(1.0332)^t$.

　　c. Decay rate $= 1 - $ decay factor $= 1 - 0.8796 = 0.1204$;
　　　growth rate $=$ growth factor $-1 = 1.0332 - 1 = 0.0332$; the sales of cassette tapes are falling by 12.04% a year, whereas the sales of CDs are rising by 3.32% a year.

　　d. $C > T \Rightarrow 865.7(1.0332)^t > 442.2(0.8796)^t \Rightarrow t \approx -4.2$ (solving graphically); the number of CDs sold overtook the number of cassette tapes sold about 4.2 years before 1990, that is, in late 1985.

23. a.

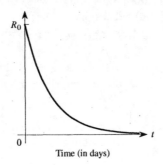

　　　Time (in days)

　　b. Using a linear estimate, we would predict a decrease of 50 grams each day, or $800 - 550 = 250$ grams. Since the graph is concave up, this is an underestimate.

　　c. A linear interpolation between $(10, 800)$ and $(30, 400)$ would predict 600 grams left after 30 days. Since the graph is concave up, the line connecting the two data points lies above the curve, so 600 grams is an overestimate.

　　d. For the scenario in part (b), since $750/800 = 0.9375$, the decay rate per day, we can estimate the amount left from 800 grams after 10 days as $800(0.9375)^{10} \approx 419.6$, or about 420 grams. For the scenario in part (c), since the change from 800 grams to 400 grams takes 20 days, this is the half-life of the substance. At 20 days, that is, 10 days after the amount is 800, the amount will be
$$800\left(\tfrac{1}{2}\right)^{1/2} = \frac{800}{\sqrt{2}} = 565.7 \text{ grams.}$$

25. The graph provided looks like the graph corresponding to a vertical shift upwards by 10 units of the reflection across the horizontal axis of the exponential function $10a^x$, where this exponential function has the value $= 5$ when $x = 1$, so $a = 0.5$. The appropriate values for A, B and C in the modified exponential expression $y = A + BC^x$ are $A = 10$, $B = -10$ and $C = 0.5$.

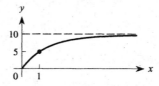

Exercising Your Algebra Skills

1. $2^m \cdot 2^n = 2^{m+n}$

2. $\dfrac{1}{2^u} \cdot \dfrac{1}{2^v} = \dfrac{1}{2^{u+v}}$

3. $5^3 \cdot 5^x = 5^{3+x}$

4. $4^{-2} \cdot 4^{3x} = 4^{-2+3x}$

5. $3^5 \cdot 3^{-2a} = 3^{5-2a}$

6. $2^{-4} \cdot 2^{-3w} = 2^{-4-3w}$

7. $\dfrac{10^{-3x}}{10^{2x}} = 10^{-3x-2x} = 10^{-5x}$

8. $\dfrac{4^{3x}}{4^{-3x}} = 4^{3x-(-3x)} = 4^{6x}$

9. $(2^x)^5 = 2^{5x}$

10. $(0.7^x)^{10} = (0.7)^{10x}$

11. $3^{x+2} = 3^x \cdot 3^2 = 9(3^x)$

12. $5^{x-2} = \dfrac{5^x}{5^2} = \dfrac{1}{25}(5^x)$

13. $10^{3x+1} = 10^{3x} \cdot 10 = 10 \cdot (10^3)^x = 10(1000^x)$

14. $\left(\dfrac{1}{2}\right)^{4x+3} = \left(\dfrac{1}{2}\right)^{4x} \cdot \left(\dfrac{1}{2}\right)^3 = \dfrac{1}{8}\left(\dfrac{1}{2}\right)^{4x} = \dfrac{1}{8}\left(\dfrac{1}{16}\right)^x$

Section 2.6   Logarithmic Functions

1. a. The graph is a straight line, which is correct since
$y = \log 10^{2x} \Rightarrow y = 2x \log 10 \Rightarrow y = 2x(1) \Rightarrow y = 2x$ (linear).

  b. The graph is a horizontal line, which is correct since
$y = \log(2x) - \log(x) \Rightarrow y = \log \dfrac{2x}{x} \Rightarrow y = \log 2 \Rightarrow y = 0.3010$ (constant).

  c. The graph is a parabola, which is correct since $y = \log 10^{x^2} \Rightarrow y = x^2$ (quadratic).

  d. The graph is a parabola, which is correct since $y = 10^{\log(x^2)} \Rightarrow y = x^2$ (quadratic). (Note: Here we ignore the fact that the function $y = 10^{\log(x^2)}$ is undefined at $x = 0$ since we want the general pattern of the functional values, and $y = 10^{\log(x^2)}$ is identical to $y = x^2$ everywhere except at $x = 0$.)

  e. The graph is a straight line, which is correct since $y = \log 3^x \Rightarrow y = x \log 3 \Rightarrow y = 0.4771x$ (linear).

  f. The graph is a straight line, which is correct since
$y = \log\left(\dfrac{10}{6^x}\right) \Rightarrow y = \log(10 \cdot 6^{-x}) \Rightarrow y = \log 10 + \log 6^{-x} \Rightarrow y = 1 - x\log 6 \Rightarrow y = 1 - 0.7782x$ (linear).

3. **a.** The exponential model for Argentina's population is $P = 34.6(1.013)^t$, where t is time measured in years after 1995.
    b. For 2005, $t = 10$, and $P = 34.6(1.013)^{10} \approx 39.4$ million.
    c. To find the doubling time, we solve $(1.013)^t = 2$ for t:
    $$\log(1.013)^t = \log 2 \Rightarrow t \cdot \log(1.013) = \log 2 \Rightarrow t = \log 2/\log(1.013) \approx 53.7 \text{ years}$$

5. The fish population is modeled by $P = P_0(0.9)^t$, where P_0 is the initial fish population and t is given in weeks. To find the half-life for this model, we solve $(0.9)^t = 0.5$ for t: $\log(0.9)^t = \log(0.5) \Rightarrow t\log(0.9) = \log(0.5) \Rightarrow t = \log 0.5/\log 0.9 \approx 6.6$ weeks.

7. **a.** A growth rate of 3% gives 1.03 as the growth factor, and $(1.03)^t = 2 \Rightarrow t = \log 2/\log 1.03 \approx 23.45$. So the approximate doubling time for 3% growth rate is 23.45 years. Similarly, for the growth rates of 4%, 5%, 6%, and 7%, the approximate doubling times are 17.61, 14.21, 11.9, and 10.24 years, respectively.
    b. The graph below is decreasing and concave up, which suggests that an exponential decay function seems to fit.

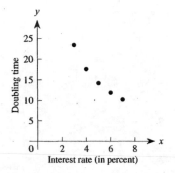

9. Letting $t = 0$ in 1990 gives data points $(0, 0.29)$ and $(12, 0.37)$; the first-class postage model can be found by solving $0.37 = 0.29c^{12}$ for c: $c = \sqrt[12]{0.37/0.29} \approx 1.02051$; so the first-class postage model is $P = 0.29(1.02051)^t$, where t is the number of years after 1990. To find when the first-class postage will reach $1.00, we solve $0.29(1.02051)^t = 1.00$ for t: $(1.02051)^t = 1.00/0.29 \Rightarrow t = \log(1.00/0.29)/\log(1.02051) \approx 61$ years; that is, in 2051.

11. For the level of carbon-14 to go down to 61% of the level in living wood, the number of half–lives t that have elapsed since the fire in question is the value t such that $0.5^t = 0.61$: $t = \log(0.61)/\log(0.5) \approx 0.71311$ half-lives.. Thus, we find that the fire took place $5730(0.71311) = 4086$ years before 1950, roughly 2137 B.C.

13. If the level of carbon-14 found in 1991 in the Shroud of Turin was 91% of the level found in newly made cloth of the same type, then the number of half–lives that have elapsed since the shroud was made is t, where $0.5^t = 0.91$: $t = \log(0.91)/\log(0.5) \approx 0.1361$, which sets the age of the Shroud in 1991 at 780 years old, or dating back to the 13th century. This is the reason why the Vatican has disclaimed the Shroud as a legitimate relic dating to the time of Jesus.

15. **a.** If the decay rate is $2\% = 0.02$, then 0.98 is the decay factor. The exponential model for the percentage of mosquitoes remaining after n months is $M = (0.98)^n$. For 1 year, $n = 12$, and $M = (0.98)^{12} \approx 0.7847$. So, about 78.5% of mosquitoes would remain after 1 year.
    b. To find how long for the population to be reduced by half, we solve $(0.98)^n = 0.5$ for n:
    $(0.98)^n = 0.5$ for n: $\log(0.98)^n = \log(0.5) \Rightarrow n\log(0.98) = \log(0.5) \Rightarrow n = \log(0.5)/\log(0.98) \approx 34$ months, or 2 years 10 months.
    c. $(0.98)^n = 0.1 \Rightarrow n = \log(0.1)/\log(0.98) \approx 114$ months, or 9.5 years.

17. The intensity, I_3, of a quake with Richter magnitude 3 is given by $I_3 = 10^3 \cdot I_0$. The intensity of a quake with Richter magnitude 6 quake is given as $I_6 = 10^6 \cdot I_0$. This means that $I_6 = 10^3 \cdot I_3$, so the quake with Richter magnitude 6 is 1,000 times as strong as the quake with magnitude 3.

19 a. $60 = 10/\log(I/I_0) \Rightarrow \log(I/I_0) = 6 \Rightarrow 10^{\log(I/I_0)} = 10^6 \Rightarrow \dfrac{I}{I_0} = 10^6 \Rightarrow I = 10^6 I_0$. So, the intensity of

   normal conversation is given by $I = 1,000,000 I_0$. Thus, conversation is a million times as loud as a threshold murmur.

   b. A noise of 150 decibels is 10^{15} times more intense than the threshold level.

   c. A noise of 120 decibels is 10^{12} times more intense than the threshold level.

   d. $d = 10\log(1,000,000 I_0 / I_0) = 10\log 10^6 = 10(6) = 60$ decibels

   e. $d = 10\log(100,000,000,000 I_0 / I_0) = 10\log 10^{11} = 10(11) = 110$ decibels

21. $(1.015)^t = (10^{\log 1.015})^t = (10^{0.0065})^t$

   $P(t) = 6(1.015)^t = 6(10^{0.0065})^t$

   $(1.015)^t = (e^{\ln 1.015})^t = e^{0.0149t}$

   $P(t) = 6(1.015)^t = 6e^{0.0149t}$

Exercising Your Algebra Skills

1. $\log x + \log x^2 + \log x^3 = \log(x \cdot x^2 \cdot x^3) = \log x^6 = 6\log x$

2. $\log x + \log \sqrt{x} = \log(x \cdot x^{1/2}) = \log x^{3/2} = \frac{3}{2}\log x$

3. $\log x^2 + \log y^3 - \log x - \log y^2 = \log\left(\dfrac{x^2}{x}\right) + \log\left(\dfrac{y^3}{y^2}\right) = \log x + \log y = \log(xy)$

4. $\log\dfrac{x}{y} - \log\dfrac{y}{x} = \log\left(\dfrac{x}{y} \div \dfrac{y}{x}\right) = \log\left(\dfrac{x}{y} \cdot \dfrac{x}{y}\right) = \log\left(\dfrac{x^2}{y^2}\right)$

5. $\log 10^{x^2} = x^2$

6. $10^{\log(x^2)} = x^2$

7. $7^x = 11 \Rightarrow x\log 7 = \log 11 \Rightarrow x = \dfrac{\log 11}{\log 7} \approx 1.23$

8. $1.05^x = 2 \Rightarrow x\log 1.05 = 2 \Rightarrow x = \dfrac{\log 2}{\log 1.05} \approx 14.2$

9. $0.4^x = 0.6 \Rightarrow x\log 0.4 = \log 0.6 \Rightarrow x = \dfrac{\log 0.6}{\log 0.4} \approx 0.56$

10. $3(1.04)^x = 5 \Rightarrow (1.04)^x = \frac{5}{3} \Rightarrow x\log 1.04 = \log(5/3) \Rightarrow x = \dfrac{\log(5/3)}{\log 1.04} \approx 13.02$

11. $12(0.86)^x = 3 \Rightarrow (0.86)^x = \frac{1}{4} \Rightarrow x\log(0.86) = \log(0.25) \Rightarrow x = \dfrac{\log(0.25)}{\log(0.86)} \approx 9.19$

12. $9(0.17)^x = 0.25 \Rightarrow (0.17)^x = \dfrac{0.25}{9} \Rightarrow x\log(0.17) = \log(0.25/9) \Rightarrow x = \dfrac{\log(0.25/9)}{\log(0.17)} \approx 2.02$

13. $4(1.05)^x = 5(1.04)^x \Rightarrow \left(\dfrac{1.05}{1.04}\right)^x = \dfrac{5}{4} \Rightarrow x\log(1.05/1.04) = \log 1.25 \Rightarrow x = \dfrac{\log 1.25}{\log(1.05/1.04)} \approx 23.32$

14. $3(0.7)^x = 6(0.5)^x \Rightarrow \left(\dfrac{0.7}{0.5}\right)^x = \dfrac{6}{3} \Rightarrow x\log(0.7/0.5) = \log(2) \Rightarrow x = \dfrac{\log(2)}{\log(0.7/0.5)} \approx 2.06$

15. $\log x = 2 \Rightarrow 10^{\log x} = 10^2 \Rightarrow x = 10^2 \Rightarrow x = 100$

16. $\log x = 0.5 \Rightarrow 10^{\log x} = 10^{0.5} \Rightarrow x \approx 3.162$

17. $\log 1,000,000 = \log 10^6 = 6$

18. $\log 0.001 = \log 10^{-3} = -3$

19. $\log(1/10,000) = \log 10^{-4} = -4$

20. $\log \sqrt{10} = \log 10^{0.5} = 0.5$

21. $\qquad \log(x^2-1) \overset{?}{=} \log(x-1)+\log(x+1) \qquad$ true

$\log[(x-1)(x+1)] = \log(x-1)+\log(x+1)$

22. $\qquad \log(x^2+1) \overset{?}{=} \log x^2 + \log 1 \qquad$ false

$\log x^2 + \log 1 = \log(x^2 \cdot 1) = \log x^2$

23. $\log\left(\dfrac{1}{x}\right) \overset{?}{=} -\log x + \log(1) \qquad$ true

$\log\left(\dfrac{1}{x}\right) = \log 1 - \log x = -\log x + \log 1$

Section 2.7   Power Functions

1.   **a.** Power function. The curve through the two points would be decreasing and concave up, so $p < 0$. Since the functional values are positive, $k > 0$.

   b. Not a power function since we have a positive functional value and a negative functional value given on the same side of the y-axis.

   c. Power function. The curve through the two points would be increasing and concave down, so $0 < p < 1$. Since the functional values are positive, $k > 0$.

   d. Power function. The curve through the two points would be increasing and concave up, so $p > 1$. Since the functional values are positive, $k > 0$.

   e. Power function. The curve is a reflection of a power curve in which the values would be decreasing and concave up, so $p < 0$. Since the functional values are negative, $k < 0$.

   f. Not a power function since we have a positive functional value and a negative functional value given on the same side of the y-axis.

3.   **a.** $f(x) = 40x^{1.05}$ is in the form $f(x) = kx^p$, so it's a power function.

   b $f(x) = 40(1.05)^x$ is in the form $f(x) = kc^x$, so it's an exponential function.

   c. $f(x) = \dfrac{1}{(1.4)^x} = \left(\dfrac{1}{1.4}\right)^x = \left(\dfrac{\frac{1}{14}}{10}\right)^x = \left(\dfrac{5}{7}\right)^x$ is in the form $f(x) = kc^x$, so it's an exponential function.

   d. $f(x) = \dfrac{-3}{x^{2.4}} = -3x^{-2.4}$ is in the form $f(x) = kx^p$, so it's a power function.

   e. $f(t) = 5t^{-3.7}$ is in the form $f(x) = kx^p$, so it's a power function.

   f. $f(q) = 1.09q - 4.37$ is in the form of a linear function, so it's neither power nor exponential.

   g. $f(t) = 12(0.35)^{-t} = 12(1/0.35)^t$ is in the form $f(x) = kc^x$, so it's an exponential function.

   h. $f(t) = 5\sqrt{t} = 5t^{-1/2}$ is in the form $f(x) = kx^p$, so it's a power function.

   i. $f(s) = \sqrt{s^2+3}$ is not in the form $f(x) = kx^p$ or $f(x) = kc^x$, so it's neither.

   j. $f(r) = \frac{4}{3}\pi r^3 = \left(\frac{4}{3}\pi\right)r^3$ is in the form $f(x) = kx^p$, so it's a power function.

   k. $f(z) = z \cdot z^{3/5} = z^{8/5}$ is in the form $f(x) = kx^p$, so it's a power function.

   l. $f(x) = x^x$ has variables for both the base and power, so it's neither power nor exponential.

m. $f(w) = w^2 3^w$ is a product of a power function and an exponential, but the product is neither power nor exponential.

n. $f(u) = 7(1.62)^{-u} = 7(1/1.62)^u$ is in the form $f(x) = kc^x$, so it's an exponential function.

5. In $g(x)$, the ratios for the x values that are 1 unit apart are essentially constant, so $g(x)$ is exponential. The corresponding ratios in $f(x)$ and $h(x)$ are not constant. So $f(x)$ and $h(x)$ are not exponential. The function f has the form $y = ax^2$: $f(x) = (16/5)x^2$, where the value $a = 16/5$ can be obtained from the equation $80 = a \cdot 5^2$, for example. The function g is exponential: $g(x) = 1.98713(1.30241)^x$, where the values $k = 1.98713$ and $c = 1.30241$ can be obtained from the equations $kc^3 = 4.39$ and $kc^{3.5} = 5.01$. The function h has the form $y = ax^3$: $h(x) = 0.4x^3$, where the value $a = 0.4$ can be obtained from the equation $10.8 = a \cdot 3^3$.

7. $W = 0.15S^{9/4} \Rightarrow 15 = 0.15S^{9/4} \Rightarrow S^{9/4} = 15/0.15 = 100 \Rightarrow S = 100^{4/9} \approx 7.7$. According to the formula, a bird weighing 15 lb needs to have a wingspan of about 7.7 ft. Since the turkey's wingspan is actually about 2.5 ft, the turkey cannot support its weight in flight.

9. $W = 0.15S^{9/4} \Rightarrow 0.15(40)^{9/4} \approx 603.6$ lb.

11. a. Data points are (1, 3) and (4, 6):
$$y = kx^p \Rightarrow 3 = k(1^p) \Rightarrow 3 = k(1) \Rightarrow k = 3;$$
$$y = 3x^p \Rightarrow 6 = 3(4)^p \Rightarrow 4^p = 2 \Rightarrow p\log 4 = \log 2 \Rightarrow p = \frac{\log 2}{\log 4} = 0.5 = \tfrac{1}{2}; \ y = 3x^{1/2}$$

b. Data points are (1, 3) and (4, 8):
$$y = kx^p \Rightarrow 3 = k(1^p) \Rightarrow 3 = k(1) \Rightarrow k = 3;$$
$$y = 3x^p \Rightarrow 8 = 3(4)^p \Rightarrow 4^p = \tfrac{8}{3} \Rightarrow p\log 4 = \log(8/3) \Rightarrow p = \frac{\log(8/3)}{\log 4} \approx 0.7075; \ y = 3x^{0.7075}$$

c. Data points are (1, 3) and (4, 10):
$$y = kx^p \Rightarrow 3 = k(1^p) \Rightarrow 3 = k(1) \Rightarrow k = 3;$$
$$y = 3x^p \Rightarrow 10 = 3(4)^p \Rightarrow 4^p = \tfrac{10}{3} \Rightarrow p\log 4 = \log(10/3) \Rightarrow p = \frac{\log(10/3)}{\log 4} \approx 0.8685; \ y = 3x^{0.8685}$$

d. Data points are (5, 20) and (6, 30):
$$y = kx^p \Rightarrow 20 = k(5^p) \text{ and } 30 = k(6)^p \Rightarrow \frac{k \cdot 6^p}{k \cdot 5^p} = \frac{30}{20} \Rightarrow \left(\frac{6}{5}\right)^p = \frac{3}{2} \Rightarrow p\log(6/5) = \log(3/2) \Rightarrow$$
$$p = \frac{\log(3/2)}{\log(6/5)} \approx 2.2239; \ k(5)^p = 20 \Rightarrow k(5^{2.2239}) = 20 \Rightarrow k \approx 0.55794; \ y = 0.55794x^{2.2239}$$

e. Data points are (1, 10) and (4, 5):
$$y = kx^p \Rightarrow 10 = k(1^p) \Rightarrow 10 = k(1) \Rightarrow k = 10;$$
$$y = 10x^p \Rightarrow 5 = 10(4)^p \Rightarrow 4^p = \tfrac{1}{2} \Rightarrow p\log 4 = \log(1/2) \Rightarrow p = \frac{\log(1/2)}{\log 4} = -\tfrac{1}{2}; \ y = 10x^{-1/2}$$

f. Data points are (2, 20) and (5, 8):
$$y = kx^p \Rightarrow 20 = k(2^p) \text{ and } 8 = k(5)^p \Rightarrow \frac{k \cdot 5^p}{k \cdot 2^p} = \frac{8}{20} \Rightarrow \left(\frac{5}{2}\right)^p = \frac{2}{5} \Rightarrow p\log(5/2) = \log(2/5) \Rightarrow$$
$$p = \frac{\log(5/2)}{\log(2/5)} = -1; \ k(2)^p = 20 \Rightarrow k(2^{-1}) = 20 \Rightarrow \frac{k}{2} = 20 \Rightarrow k = 40; \ y = 40x^{-1}$$

13. a. Let $v = 10$: $P = 0.015v^3 \Rightarrow P = 0.015(10^3) = 15$ watts.
b. Let $v = 20$: $P = 0.015v^3 \Rightarrow P = 0.015(20^3) = 120$ watts.
c. The power is increased by a factor of $2^3 = 8$ since $(20)^3 = (2 \cdot 10)^3 = 2^3 \cdot 10^3$. Note that $15 \cdot 8 = 120$.
d. $P = 0.015(5^3) = 1.875$ watts and $P = 0.015(10^3) = 15$ watts; $\dfrac{15}{1.875} = 8$; yes because, in general,
$$\frac{0.015(2v)^3}{0.015(v)^3} = \frac{8v^3}{v^3} = 8.$$

  e. The amount of power generated by a windmill with wind speed of 12 mph is $P = 0.015(12^3) =$
  25.92 watts. For 250 kilowatts $= 250,000$ watts, the community would need
  $250,000/25.92 \approx 9645$ windmills.

  f. $P = 0.015v^3 \Rightarrow 100 = 0.015v^3 \Rightarrow v^3 = 100/0.015 \Rightarrow v = \sqrt[3]{100/0.015} \approx 19$ mph.

15. **a.** A range of $0 \le y \le 0.7$ should produce a satisfactory plot.
  b. A range of $0 \le y \le 1.5$ should produce a satisfactory plot.
  c. A range of $0 \le y \le 4$ should produce a satisfactory plot.

  d. Only the function $y = x^{1/3}$ is defined for negative x-values. So, the screen shows on the left side of the
  x-axis just the graph of $y = x^{1/3}$.

17. Letting $t = 1$ in 1990, we have the data points (1, 442.2) and (9, 158.5) for cassette tapes and (1, 865.7) and
  (9, 1124.3) for CDs.

  a. $T = 442.2t^p \Rightarrow 158.5 = 442.2(9)^p \Rightarrow p = \log(158.5/442.2)/\log 9 \approx -0.467$. So, $T = 442.2t^{-0.467}$,
  where $t = 1$ in 1990.

  b. $C = 865.7t^p \Rightarrow 1124.3 = 865.7(9^p) \Rightarrow p = \log(1124.3/865.7)/\log 9 \approx 0.119$. So, $C = 865.7t^{0.119}$,
  where $t = 1$ in 1990.

  c. Using a grapher, we graph $y_1 = 442.2x^{-0.467}$ and $y_2 = 865.7x^{0.119}$, and we find that the graphs intersect
  at about $x = 0.32$, which means that $y_2 > y_1$ when $x > 0.32$. Alternatively, solving algebraically, we
  get:

$$865.7t^{0.119} = 442.2t^{-0.467} \Rightarrow \frac{t^{0.119}}{t^{-0.467}} = \frac{442.2}{865.7} \Rightarrow t^{0.586} = 442.2/865.7 \Rightarrow t = \sqrt[0.586]{442.2/865.7} \approx 0.32.$$

  Since $t = 1$ represents 1990, then $t = 0.32$ represents early 1989, so the CDs sales overtook the cassette
  tapes sales in early 1989.

19. $H = 1250$ ft $= 1250/5280$ mi ≈ 0.2367 mi. Using the formula found in Problem 18, we get:

  $D = \sqrt{(0.2367)^2 + 7920(0.2367)} \approx 43$ mi.

21. From Problem 18, we add $\frac{1}{20}$ mile to the tower height to get a new value for H: $H = 0.275 + 0.05 =$
  0.325 mi; $D = \sqrt{(0.325)^2 + 7920(0.325)} \approx 50.74$ mi. This gives the transmitter a range of only an
  additional 4 miles, less than a 9% increase. However, the difference in service areas is
  $\pi(50.74^2) - \pi(46.67^2) \approx 1246$ mi, an 18% increase.

23. A satellite in geosynchronous orbit at an altitude of 23,000 miles will sweep out a circle of radius
  $(23,000 + 3960)$ miles whose center is the center of Earth during every twenty-four hour day. Thus, its
  orbital velocity must be $(2\pi \cdot 26,960)/24 \approx 7058$ mph.

25.

H	$D = 89\sqrt{H}$	$D(H) = \sqrt{H^2 + 7920H}$
0.1 mile	28.14427	28.1427
1 mile	89	89
10 miles	281.44	281.60
100 miles	890	895.54

27. For the points P (0, 0), Q(1, 1) and R(2, 4) on the graph of the function $f(x) = x^2$, the slope of $\overline{PQ} = 1$, the
  slope of $\overline{QR} = 3$, and the slope of $\overline{PR} = 2$.

29. The points $P(a, a^2)$, $Q(a+h, (a+h)^2)$ and $R(a+2h, (a+2h)^2)$ on the graph of the function $f(x) = x^2$ determine the segments PQ, PR, and QR:

$$m_{\overline{PQ}} = \frac{(a+h)^2 - a^2}{a+h-a} = \frac{a^2 + 2ah + h^2 - a^2}{h} = \frac{h(2a+h)}{h} = 2a+h$$

$$m_{\overline{PR}} = \frac{(a+2h)^2 - a^2}{a+2h-a} = \frac{a^2 + 4ah + 4h^2 - a^2}{2h} = \frac{4h(a+h)}{2h} = 2(a+h) = 2a+2h$$

$$m_{\overline{QR}} = \frac{(a+2h)^2 - (a+h)^2}{a+2h-(a+h)} = \frac{a^2 + 4ah + 4h^2 - a^2 - 2ah - h^2}{a+2h-a-h} = \frac{2ah + 3h^2}{h} = \frac{h(2a+3h)}{h} = 2a+3h$$

average of $m_{\overline{PQ}}$ and $m_{\overline{QR}} = \frac{m_{\overline{PQ}} + m_{\overline{QR}}}{2} = \frac{(2a+h) + (2a+3h)}{2} = \frac{4a + 4h}{2} = 2a + 2h = m_{\overline{PR}}$

31. The capacities of different calculators will differ. Therefore, the highest power of 10 and the smallest positive number that can be represented vary from one to another calculator.

Exercising Your Algebra Skills

1. $9^{1/2} = \sqrt{9} = 3$
    2. $9^{-1/2} = \frac{1}{9^{1/2}} = \frac{1}{\sqrt{9}} = \frac{1}{3}$
    3. $8^{4/3} = (8^{1/3})^4 = \left(\sqrt[3]{8}\right)^4 = 2^4 = 16$

4. $8^{-4/3} = \frac{1}{8^{4/3}} = \frac{1}{16}$
    5. $x^4 \cdot x^3 = x^7$
    6. $x^6 \cdot x^{-8} = x^{-2}$

7. $\frac{r^8}{r^4} = r^4$
    8. $\frac{z^{12}}{z^{-9}} = z^{12+9} = z^{21}$
    9. $x^{3/4} \cdot x^{5/4} = x^{8/4} = x^2$

10. $a^{-2/3} \cdot a^{5/3} = a^{-2/3+5/3} = a^{3/3} = a$
    11. $\frac{x^{3/4}}{x^{5/4}} = x^{3/4-5/4} = x^{-2/4} = x^{-1/2}$

12. $\frac{a^{-2/3}}{a^{5/3}} = a^{-2/3-5/3} = a^{-7/3} = \frac{1}{a^{7/3}}$
    13. $\frac{a^{5/3}}{a^{-2/3}} = a^{5/3-(-2/3)} = a^{7/3}$

Section 2.8   Comparing Rates of Growth and Decay

1.   **a.** Use the estimated window $0 \le x \le 3$ and $0 \le y \le 6$; the intersection point is at about $(1.4, 2.6)$.
    b. Use the estimated window $1.4 \le x \le 8$ and $2.6 \le y \le 500$.
    c. Use the estimated window $0 \le x \le 15$ and $0 \le y \le 15,000$; the intersection point is at about $(9.9, 982)$.

3. The graph of $y = 0.6^x$ overtakes $y = x^{-6}$ at $x \approx 1.098$, as they both decay to zero.

5.   **a.** For $f(x) = 4x - 9$ and $\Delta x = 5 - 2 = 3$:
$$\frac{f(5) - f(2)}{\Delta x} = \frac{11 - (-1)}{3} = \frac{12}{3} = 4$$
    b. For $f(x) = 4x - 9$ and $\Delta x = 3 - (-2) = 5$:
$$\frac{f(3) - f(-2)}{\Delta x} = \frac{3 - (-17)}{5} = \frac{20}{5} = 4$$
    c. For $f(x) = -3x + 4$ and $\Delta x = 5 - 2 = 3$:
$$\frac{f(5) - f(2)}{\Delta x} = \frac{11 - (-2)}{3} = \frac{-9}{3} = -3$$

d. For $f(x) = -3x + 4$ and $\Delta x = 3 - (-2) = 5$:

$$\frac{f(3) - f(-2)}{\Delta x} = \frac{-5 - 10}{5} = \frac{-15}{5} = -3$$

e. $4x - 3y = 12 \Rightarrow y = \frac{4}{3}x - 4 \Rightarrow f(x) = \frac{4}{3}x - 4$ and $\Delta x = 5 - 2 = 3$

$$\frac{f(5) - f(2)}{\Delta x} = \frac{(8/3) - (-4/3)}{3} = \frac{12/3}{3} = \frac{4}{3}$$

7. For $f(x) = x^2$:

a. $\Delta x = 1 - 0 = 1$ $\dfrac{f(1) - f(0)}{\Delta x} = \dfrac{1 - 0}{1} = 1$

b. $\Delta x = 2 - 0 = 2$ $\dfrac{f(2) - f(0)}{\Delta x} = \dfrac{4 - 0}{2} = 2$

c. $\Delta x = 2 - 1 = 1$ $\dfrac{f(2) - f(1)}{\Delta x} = \dfrac{4 - 1}{1} = 3$

9. **a.** For $f(x) = \sqrt{x}$, the numerical order from the smallest to the largest average rate of change would be the intervals from 1 to 2, next from 0 to 2, and then from 0 to 1.

b. $f(x) = \sqrt{x}$, $\Delta x = 1 - 0 = 1$ $\dfrac{f(1) - f(0)}{\Delta x} = \dfrac{1 - 0}{1} = 1$

c. $f(x) = \sqrt{x}$, $\Delta x = 2 - 1 = 1$ $\dfrac{f(2) - f(1)}{\Delta x} = \dfrac{\sqrt{2} - 1}{1} = \dfrac{1.414 - 1}{1} \approx 0.414$

b. $f(x) = \sqrt{x}$, $\Delta x = 2 - 0 = 2$ $\dfrac{f(2) - f(0)}{\Delta x} = \dfrac{\sqrt{2} - 0}{2} = \dfrac{\sqrt{2}}{2} \approx 0.707$

11. **a.** $\overline{PQ}, \overline{PR}, \overline{QR}$.. The slope is nonnegative in all three cases. The lowest value of the slope is the slope of segment PQ; the highest value of the slope is the one of segment QR. The slope of segment PR is an intermediate value.

b. If the function is increasing and concave down, the list in the order of increasing slope is $\overline{QR}, \overline{PR}, \overline{PQ}$, that is the reverse of the previous case.

Section 2.9   Inverse Functions

1. **a.** As long as the pool does not fill up and overflow, the height of water in the pool will be an increasing function and thus have an inverse. The inverse function tells you how long you need to run the water in order to fill the pool to a desired height.

b. The distance from New York is an increasing function, and thus has an inverse. The inverse function tells us how long one will be in the air before reaching a given distance away from New York.

c. For small classes this function is likely (though by no means guaranteed unless the class has only one student) to have an inverse. For large classes there will certainly be two students that have the same height, so that the function will not have an inverse function.

d. Except in rather bizarre circumstances, there will most likely be two different customers that pay the same amount for lunch. Thus, the function does not have an inverse.

e. The length of your fingernail will be an increasing function until you cut the nail. Thus, for the interval between manicures the function will have an inverse. The information carried by the inverse function is the time it takes for you fingernail to grow to a given length.

f. What with the great variation in the heating characteristics of houses and the habits of families, $f(T)$ is *very unlikely* to be a function at all. Moreover, for a large community it is likely that there are two families that have the same monthly bill but set their thermostats at different temperatures. Thus the function will usually fail to have an inverse.

g. Since the snow will partially melt (or at least compact) between snowfalls, the level will go up and down. Thus, the function does not have an inverse.

h. Although the amount of snow continues to increase over the season, never does it snow continuously, not even in Buffalo, NY. Thus, there will be intervals during which the depth amount of snow is constant. Due to these intervals, the function does not have an inverse.

3. a. The domain of f is the collection of values for which f is uniquely and unambiguously defined. Recording the domain of f as a set we get $\{0, 1, 2, 3, 4, 5\}$. The range of f is the collection of values assumed by f. Recording the range as a set, we get $\{1.12, 1.44, 1.84, 2.05, 2.48, 2.94\}$.

  b. The function f^{-1} is given by the following table of values

x	1.12	1.44	1.84	2.05	2.48	2.94
$f^{-1}(x)$	5	4	3	2	1	0

5. a. Each speed that a Trans Am can obtain corresponds to an unique time elapse (that is, the time it would take to reach that speed when starting from a standstill). So the data represents a function. The speed of a Trans Am will increase smoothly and continually until it tops out. From the table, we expect that the speed would be considerably above 70 mph. Increasing functions always have inverses, so, at least on the interval under consideration, the function $v = f(T)$ has an inverse. The inverse function tells us for any time elapse of a Trans Am starting from a standstill how fast the car is finally moving.

  b. The inverse function tells us the speed a Trans Am will reach after T seconds of acceleration. The value $f^{-1}(5.52) = 50$ mph can simply be read off the table. An estimate of $f^{-1}(7) \approx 58$ mph can be had by observing that $f^{-1}(7)$ will lie between $f^{-1}(5.52)$ and $f^{-1}(7.38)$ for which values of 50 mph and 60 mph respectively can be read from the table. We estimate the value of $f^{-1}(7)$ with the following reasoning. The time of 7 seconds is part way between 5.52 seconds and 7.38 seconds, in fact it is $(7 - 5.52)/(7.38 - 5.52) = 1.48/1,86 \approx 0.8$ of the way. So it is reasonable to expect (reasonable at least in the absence of other information) that the speed achieved in 7 seconds will be 0.8 of the way between 50 mph and 60 mph, or $0.8(60 - 50) = 8$ mph from 50 mph. , Hence, we get the estimate of 58 mph.

7. $p = (1.04)^t \Rightarrow \log p = t \log 1.04 \Rightarrow t = \dfrac{\log p}{\log 1.04}$; interchanging the role of t and p to use the same independent variable, we get $p = \dfrac{\log t}{\log 1.04}$. So, $p^{-1}(t) = \left(\dfrac{1}{\log 1.04}\right)\log t$, or $p^{-1}(t) \approx 58.7084 \log t$.

9. The choice of the Fahrenheit scale would give a more accurate reading, because it is a "tighter" scale. That is, each degree measures a smaller change in temperature than does a Celsius degree.

11. Since in the graph of an increasing function, larger x-values always yield larger y-values. Conversely an increased y-value must imply an increased x-value, forcing the inverse function to be increasing.

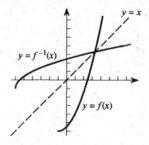

As one can see from by reflecting the graph about the line $y = x$ in the preceding graph , $f^{-1}(x)$ is concave down. Reflecting a function that is increasing and concave down across the line $y = x$, of course, gives a function that is increasing and concave up.

13. A sketch of the graph of f^{-1} is as follows:

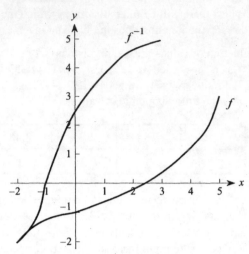

15. **a.** $P(t) = 80(0.75)^t \Rightarrow P = 80(0.75)^t \Rightarrow \log P = \log 80 + t \log 0.75 \Rightarrow t = \dfrac{\log P - \log 80}{\log 0.75} \Rightarrow$

$$t = f^{-1}(P) = \dfrac{\log P - 1.903}{-0.1249}$$

b. $t = \dfrac{\log 25 - 1.903}{-0.1249} \approx 4$ days

17. **a.** Each letter of the alphabet is matched with only one letter of the alphabet; that is, since each correspondence is unique, the assignment is uniquely reversible.

  b. IS THIS MATH?

Exercising Your Algebra Skills

1. $c^{25} = 14 \Rightarrow c = \sqrt[25]{14} \approx 1.113$

2. $0.07^t = 3 \Rightarrow t \log 0.07 = \log 3 \Rightarrow t = \log 3 / \log 0.07 \approx -0.4131$

3. $0.84^k = 0.20 \Rightarrow k \log 0.84 = \log 0.20 \Rightarrow k = \log 0.20 / \log 0.84 \approx 9.2309$

4. $m^{1995} = 4 \Rightarrow m = \sqrt[1995]{4} \approx 1.0007$

5. $17b^8 = 32 \Rightarrow b^8 = 32/17 \Rightarrow b = \sqrt[8]{32/17} \approx 1.0823$

6. $25c^9 = 8 \Rightarrow c^9 = 8/25 \Rightarrow c = \sqrt[9]{8/25} \approx 0.8811$

7. $4(1.02)^x = 7 \Rightarrow (1.02)^x = 7/4 \Rightarrow x \log 1.02 = \log(7/4) \Rightarrow x = \log(1.75) / \log 1.02 \approx 28.2597$

8. $2(0.75)^t = 1 \Rightarrow (0.75)^t = 0.5 \Rightarrow t \log 0.75 = \log 0.5 \Rightarrow t = \log 0.5 / 0.75 \approx 2.4094$

9. $y = 12x^{7/2} \Rightarrow x^{7/2} = y/12 \Rightarrow x = (y/12)^{2/7}$

10. $y = 12(1.06)^t \Rightarrow t \log 1.06 = \log(y/12) \Rightarrow t = \dfrac{\log(y/12)}{\log 1.06}$

11. $Q = 27w^{-3/4} \Rightarrow w^{-3/4} = \dfrac{Q}{27} \Rightarrow w = \left(\dfrac{Q}{27}\right)^{-4/3} = \left(\dfrac{27}{Q}\right)^{4/3} = \dfrac{81}{Q^{4/3}}$

12. $L = 125(0.92)^t \Rightarrow (0.92)^t = L/125 \Rightarrow t \log(0.92) = \log(L/125) \Rightarrow t = \dfrac{\log(L/125)}{\log(0.092)}$

13. $F = ma \Rightarrow a = \dfrac{F}{m}$

14. $E = mc^2 \Rightarrow m = \dfrac{E}{c^2}$

15. $P = kVT \Rightarrow V = \dfrac{P}{kT}$

16. $K = \tfrac{1}{2}mv^2 \Rightarrow v^2 = \dfrac{2k}{m} \Rightarrow v = \sqrt{\dfrac{2k}{m}}$

Chapter 2 Review Problems

1.   **a.** Exponential; base $c < 1$
    b. Power function; $0 < p < 1$
    c. Exponential; base $c > 1$

2.   **a.** Exponential; decreasing, concave up, doesn't go through the origin; with domain being all real numbers.
    b. Power; increasing, concave up, goes through the origin.
    c. Logarithmic; increasing, concave down; with domain $x > 0$.
    d. Power; decreasing, concave up; with the positive y-axis and the positive x-axis as asymptotes.
    e. Power; increasing, concave down, goes through the origin.
    f. Exponential; increasing, concave up, doesn't go through the origin, with domain being all real numbers.
    g. Exponential; of the form $y = kc^x$
    h. Power; of the form $y = kx^p$
    i. Exponential; of the form $y = kc^x$
    j. Power; of the form $y = kx^p$
    k. Power; of the form $y = kx^p$ since $y = 1/\sqrt{x} \Rightarrow y = x^{-1/2}$.
    l. Linear; of the form $y = mx + b$ since $5x - 3y = 15 \Rightarrow y = \tfrac{5}{3}x - 5$.

    m. Linear, because the differences between the successive y values are constant.
    n. Exponential, because the ratios of the successive y values are constant.

3.   **a.** A   **b.** D   **c.** C   **d.** B   **e.** E   **f.** I
    g. H   **h.** G   **i.** F   **j.** J   **k.** L   **l.** K

4.   **a.** (6)   **b.** (5)   **c.** (4)   **d.** (3)   **e.** (1)

5.   **a.** The scale for the vertical axis is not uniform — the distances between 0, 1, 5, and 20 are the same — so this data appears linear even though it is not.

    b.

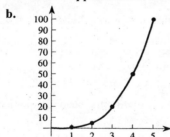

    c. Since the function is increasing and concave up and assuming that the point (0, 0) is part of the data (as the graph in the book indicates), a possible function that might be an appropriate model would be a power function. Also, since the ratios of successive functional values are not constant, an exponential function cannot be a possible model.

6. With data points (0, 5) and (3, 8.5), we have
$$y = kc^t \Rightarrow 5 = kc^0 \Rightarrow k = 5; \; y = 5c^t \Rightarrow 8.5 = 5c^3 \Rightarrow c^3 = 1.7 \Rightarrow c = \sqrt[3]{1.7} \Rightarrow c \approx 1.1935; \; \text{so}$$
$y = 5(1.1935)^t$, where the growth factor is 1.1935.

7. **a.** $P(t) = 0.1t + 1.5$, where t is the number of years since 1990.

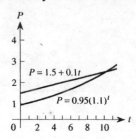

$P(t) = 0.1t + 1.5 \Rightarrow 0.1t + 1.5 = 2 \Rightarrow 0.1t = 0.5 \Rightarrow t = 5;$ the profits first exceed $2 million in 1995.

b. If the profits of Ord grow at the rate of 10% each year and they were $950,000 in 1990 (the initial profit value), then the equation for the Ord profit function is exponential with growth factor $c = 1.1$:
$P(t) = 950,000(1.1)^t$. The profits exceed $2 million when $950,000(1.1)^t > 2,000,000$, or in late 1997.

c. We solve graphically the inequality $0.95(1.1)^t > 0.1t + 1.5$, which gives $t > 10.26$. So, the profits of Ord exceed those of Alamo during 2000.

8. **a.** Letting $t = 0$ in 1967, we have the data points (0, 800) and (27, 8000);
$k = 800$ and $y = 800c^t \Rightarrow 8000 = 800c^{27} \Rightarrow c^{27} = 10 \Rightarrow c = \sqrt[27]{10} \approx 1.089$. So the exponential function that models the bald eagle population is $P(t) = 800(1.089)^t$. To find the doubling time, we solve the equation $(1.089)^t = 2$ for t: $\; t \log 1.089 = \log 2 \Rightarrow t = \log 2 / \log 1.089 \approx 8.1$ years.

b. Since $t = 38$ in 2005, $P(38) = 800(1.089)^{38} \approx 20,424$ eagles.

c. $800(1.089)^t = 20,000 \Rightarrow (1.089)^t = 20,000/800 = 25 \Rightarrow t \log(1.089) = \log 25 \Rightarrow t = \log 25 / \log 1.089 \approx 37.7$ years. So, the eagle population can be expected to reach 20,000 during 2004.

9. Since $F(30) = 48$, we have:
$$F(t) = 172(1 + a)^t + 40 \Rightarrow 172(1 + a)^{30} + 40 = 148 \Rightarrow (1 + a)^{30} = 108/172 \Rightarrow a = \sqrt[30]{108/172} - 1 \approx -0.015392.$$
So, $F(180) = 172(1 - 0.015392)^{180} + 40 \approx 50.50$.

10. Using the black thread method, we estimate that an equation that fits the data is $W \approx 2t + 141$. If we let $t = 0$ in 1980, then 2005 corresponds to $t = 25$. So, $W \approx 2(25) + 141 = 191$ million.

11. Using the black thread method, we estimate that an equation that fits the data is $D \approx 382t + 2048$; the slope of the line is 382, which tells us that the number of deaths from accidental poisoning is increasing by 382 deaths per year. If $t = 0$ in 1980, then 2003 corresponds to $t = 23$.
So, $D \approx 382(23) + 2048 \approx 10,834$ deaths.

12. **a.** This graph corresponds to an exponential function $F(x) = kc^x$. Since we know the value of the function when $x = 0$, we get $k = 8$. Since $F(-3) = 21$, we have
$8c^{-3} = 21 \Rightarrow c^{-3} = (21/8) \Rightarrow c^3 = (8/21) \Rightarrow c = \sqrt[3]{8/21} \approx 0.72492$. So, $F(x) \approx 8(0.72492)^x$.

b. Since the graph is a line, we use the point-slope form with $(-3, 21)$ and $(0, 8)$:
$m = (21 - 8)/-3 - 0 = -\frac{13}{3}$ and $y - 8 = -\frac{13}{3}(x - 0) \Rightarrow y = -\frac{13}{3}x + 8;$ that is, $G(x) = -\frac{13}{3}x + 8$.

c. Using (2, 7) and (4, 18), we get $H(x) = kx^x \Rightarrow 7 = kc^2$ and $18 = kc^4$; dividing the second equation by the first equation, we get: $\dfrac{kc^4}{kc^2} = \dfrac{18}{7} \Rightarrow c^2 = \dfrac{18}{7} \Rightarrow c = \sqrt{\dfrac{18}{7}} \approx 1.60357$.

So, $k = 7/c^2 = 7/(1.60357)^2 \approx 2.7222$. Thus, $H(x) \approx 2.7222(1.60357)^x$.

13. **a.** Since $t = 0$ in 1990, we have (0, 1055) and (5, 1573) as data points. To construct a linear function, we first find the slope: $m = (1573 - 1055)/(5 - 0) = 518/5 = 103.6$; so, $y - 1055 = 103.6(t - 0) \Rightarrow$ $y = f(t) = 103.6t + 1055$.

    b. Since $t = 13$ in 2003, $f(13) = 103.6(13) + 1055 \approx \$2401.8$ billion.

    c. Using (0, 1055) and (5, 1573) to find an exponential model, we have $y = kc^t \Rightarrow$ $1055 = kc^0 \Rightarrow k = 1055$ and $y = 1055c^t \Rightarrow 1573 = 1055c^5 \Rightarrow c = \sqrt[5]{1573/1055} \approx 1.0832$. So, the exponential function is $F(t) \approx 1055(1.0832)^t$.

    d. Since $t = 13$ in 2003, $F(13) = 1055(1.0832)^{13} \approx \$2981.7$ billion.

    e. For $t = 20$, linear: $F(20) = 103.6(20) + 1055 \approx \$3127$ billion; exponential: $F(20) = 1055(1.0832)^{20} \approx \$5217$ billion.

14. Let $t =$ number of years since 1980 and $I =$ median family income in thousands of dollars.

    a. Since $t = 0$ in 1980, we have (0, 17.7) and (17, 37) as data points; $m = (37 - 17.7)/17 \approx 1.135$; so, $I = 1.135t + 17.7$, or $I = 1135t + 17,700$.

    b. $I = I_0 c^t \Rightarrow 17.7 = I_0 c^0 \Rightarrow I_0 = 17.7$; $I = 17.7c^t \Rightarrow 37 = 17.7c^{17} \Rightarrow c = \sqrt[17]{37/17.7} \approx 1.044$; so, $I = 17.7(1.044)^t$, or $I = 17,700(1.044)^t$.

t	Linear Model	Exponential Model
0	$17,700	$17,700
5	23,375	21,952
10	29,050	27,226
15	34,725	33,766
17	37,000	37,000
25	46,075	51,938
30	51,750	64,415

    c.
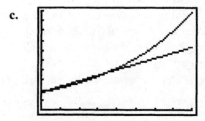

    The variable t ranges from 0 to 40.

    d. Linear: $I = 1135(23) + 17,700 \approx \$43,805$
    Exponential: $I = 17,700(1.044)^{23} \approx \$47,652$

    e. Power function.

15. **a.** Since the linear model predicted $29,050 and the exponential model predicted $27,226, the linear model comes closer to the actual value $28,900, and so it seems more accurate.

    b. The three data points seem to fall in a line. So, the shape suggests the data behaves linearly.

16. a. 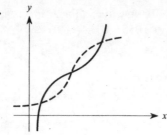   **b.**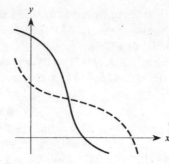

    c. The inverse function does not exist since the function is not monotonic.

17. a. The domain of $f(t) = 0.5\log(2t-4)$ is the set of values of t for which $2t-4 > 0$; that is $t > 2$; the inverse function is given by:

$$y = 0.5\log(2t-4) \Rightarrow 2y = \log(2t-4) \Rightarrow 10^{2y} = 10^{\log(2t-4)} \Rightarrow (10^2)^y = 2t-4 \Rightarrow 100^y - 2t - 4 \Rightarrow$$

$$2t = 100^y + 4 \Rightarrow t = \frac{100^y + 4}{2};$$ interchanging the roles of t and y to use the same independent variable,

we get $y = f^{-1}(t) = \dfrac{100^t + 4}{2}$.

    b. $y = x^3 + 6 \Rightarrow x^3 = y - 6 \Rightarrow x = \sqrt[3]{y-6}$; interchanging the roles of y and x, we get:

$y = g^{-1}(x) = \sqrt[3]{x-6}$.

18. Since the values of x are uniformly spaced, but the values of $F(x)$ are not, the function is not linear. Since the x values are 1 unit apart and the ratios of the given successive $F(x)$ values are $9/6 = 1.5$, $13.5/9 = 1.5$, $20.25/13.5 = 1.5$, and $30.375/20.25 = 1.5$, the function is exponential with growth factor $c = 1.5$; using the data point $(1, 6)$ in the equation $F(x) = k(1.5)^x$, we get $F(1) = k(1.5)^1 = 6 \Rightarrow k = 6/1.5 \Rightarrow k = 4$; so the formula for F is $F(x) = 4(1.5)^x$.

19. The function g is increasing and linear. The function h is increasing and concave down. Since the values of the function increase more slowly as the values of x increase. The function f is increasing and concave up. Since the values of the function increase faster as the values of x increase.

20. a. (i) For convenience, we take $t = 0$ to represent the year 1960. The value of the constant k in the equation $f(t) = kc^t$ is $k = 0.2$ (the cost of an ice cream cone in 1960). Since $t = 40$ in 2000 and we know that $(40, 1.80)$ is also a data point, we have $f(t) = 0.2c^t \Rightarrow f(40) = 0.2c^{40} = 1.8 \Rightarrow c^{40} = 1.8/0.2 \Rightarrow c = \sqrt[40]{9} \approx 1.0565$; so, the function is $f(t) = 0.2(1.0565)^t$, where $t = 0$ in 1960.
(ii) In 2005, $t = 45$, so $f(45) = 0.2(1.0565)^{45} \approx 2.37$. The price of an ice cream cone in 2005 will be about \$2.37.

    b. Using $(0, 2)$ and $(40, 9)$, we get $k = 2$ and $f(40) = 2c^{40} = 9 \Rightarrow c^{40} = 4.5 \Rightarrow c = \sqrt[40]{4.5} \approx 1.0383$. The average price of a movie ticket has a growth rate of $1.0383 - 1 = 0.0383$, or 3.83%, whereas the price of an ice cream cone has a growth rate of $1.0565 - 1 = 0.0565$, or 5.65%; so the price of ice cream is growing faster.

    c. To find when the price of ice cream will be the same as the price of a movie ticket, we solve the equation $0.2(1.0565)^t = 2(1.0383)^t$ for t:

$$0.2(1.0565)^t = 2(1.0383)^t \Rightarrow \left(\frac{1.0565}{1.0383}\right)^t = \frac{2}{0.2} = 10 \Rightarrow t\log(1.0565/1.0383) = \log 10 \Rightarrow$$

$t = \dfrac{\log 10}{\log(1.0565/1.0383)} \approx 132.509$ years. Since t is the number of years since 1960, the prices will be the same during the year 2093.

    d. From part (c), $t = 132.509$, so $f(132.509) = 2(1.0383)^{132.509} \approx 291.04$. In 2093, the average price of a movie ticket will be \$291.04.

    e. No, because the year 2093 is too far away from the year 1960 and the model may not follow the actual trend of the prices that far into the future.

21.   **a.** If the number of students decreases by $10\% = 0.1$ per year, the decay factor is $1 - 0.1 = 0.9$. At $t = 0$, 1600 students were enrolled. So, the model function is $f(t) = kc^t \Rightarrow f(t) = 1600(0.9)^t$. For $t = 3$, we have $f(t) = 1600(0.9)^3 = 1166.4$. Thus, about 1166 students will be enrolled in remedial English in 3 years.

  b. To find when the number of students is 15, we solve the equation $1600(0.9)^t = 15$ for t:

  $$(0.9)^t = 15/1600 \Rightarrow t\log(0.9) = \log(15/1600) \Rightarrow t = \frac{\log(15/1600)}{\log(0.9)} \approx 44.325.$$ It will take about 45 years

  for the number of students enrolled in remedial English to be reduced to 15 students.

22.   If the level of a drug decreases at a rate of $30\% = 0.3$, then the decay factor is $1 - 0.3 = 0.7$. Since the given initial dose is 150 mg, we have the equation $f(t) = 150(0.7)^t$; to find how long it takes to bring the drug level down to under 20 mg, we solve the equation $150(0.7)^t = 20$ for t:

  $$(0.7)^t = 2/15 \Rightarrow t\log(0.7) = \log(2/15) \Rightarrow t = \frac{\log(2/15)}{\log(0.7)} \approx 5.6 \text{ hours};$$

  5% of the original level would be $(0.05)(150) = 7.5$ mg; to find how long it takes to bring the drug level down to 7.5 mg, we solve the equation $150(0.7)^t = 7.5$ for t: $t = \dfrac{\log(7.5/150)}{\log(0.7)} \approx 8.4$ hours.

Chapter 3   Fitting Functions to Data

Section 3.2   Linear Regression Analysis

1. a. (iii)
    b. (i)
    c. (iv)
    d. (ii)
    e. (v)

3. a. A correlation coefficient is always between −1 and 1.
    b. Slope and correlation coefficient must have the same sign.
    c. The smaller x is, the larger y will be.

5. The linear correlation, r, is 0.990, which exceeds 0.811, the conventional critical value for samples of size 6. The equation of best fit is *Time* = 0.231 • *Speed* − 4.06 which predicts that it will take 6.3 seconds to accelerate to 45 mph and 16.7 seconds to reach 90 mph. Interpolations are inherently more reliable than extrapolations; thus the prediction for 45 mph is more likely to be accurate than the prediction for 90 mph.

7. The correlation remains the same as in Problem 6, which is 0.992. The new regression equation is $P = 0.254t - 10.682$, so the constant term changes but the linear coefficient remains the same. On the new time scale, $t = 0$ corresponds to 1900. Using 0, 5, 10, etc., for the years, would again change the constant term, giving the equation $P = 0.254t + 4.575$.

9. a. The new equation is $H = 1.78S + 50.863$.
    b. The correlation is the same as we obtained in Example 5: 0.95.
    c. Solving $S = 0.51H - 25.016$ for H we get $H = 1.96S + 49.051$.
    d. As noted in the hint, the errors are weighted differently when we take S as the independent variable than they are when H is the independent variable.

11. a.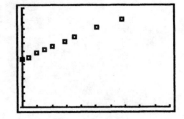

    $r = 0.999$; the scatter plot and correlation coefficient both indicate a strong linear relationship between temperature and solubility.
    b. $S = 0.871T + 67.508$, where T is temperature and S is solubility.
    c. $0.871(40) + 67.508 = 102.348$, so the model predicts a solubility of 102.3.
    d. At −5°C, the solvent might be frozen.

13. The equation of the best fit line is $v = -964.97t - 0.32$; the units for the slope are cm/sec/sec or cm/sec².

15. a. A reasonable choice is to let mass be the independent variable and length the dependent variable.
    b. $r = 0.99998$, which indicates significant correlation.
    c. $L = 0.0404m - 0.01$
    d. The units of the slope are centimeters per gram. The slope of 0.0404 indicates that the length increases by about 0.04 cm for each additional gram of mass.
    e. $L = 0.0404m$

17. **a.** $A = 28.23t + 963.82$, where $t = 0$ in 1965.

   b. Using 45 for t, we get a predicted consumption of $28.23(45) + 963.82 = 2234.170$ million tons.

   c. Solving $2000 = 28.23t + 963.82$ for t yields $t = 36.70$, so we predicts a consumption of 2000 million tons late in 2001.

19. **a.** $C = 0.022t + 13.707$, where C is cost in dollars and t is the number of free minutes.

   b. The slope indicates that the cost per free minute is about 2.2 cents. The vertical intercept indicates a fixed cost for the service of about \$13.71.

   c. $0.022(2000) + 13.707 = 57.707$, so the model predicts a cost of \$57.71 for 2000 free minutes.

21. **a.** The y height of the line that corresponds to each x-coordinate is found by substituting the x-value into the equation. For example, the height at $x = 1$ is $20(1) + 30 = 50$. The sum of squared vertical distances between the points and the line $y = 20x + 30$ is thus

$$(50 - 48)^2 + (70 - 68)^2 + (90 - 93)^2 + (110 - 114)^2 = 33.$$

   b. We want to increase the slope and decrease the vertical intercept, that is, rotate the line a bit counterclockwise around the center of the data.

Section 3.3   Fitting Nonlinear Functions to Data

1. i. Exponential
   ii. None
   iii. Power
   iv. Exponential
   v. Power
   vi. None
   vii. Logarithmic
   viii. Power

3. **a.** The model in Example 1 is $P = 3.069(1.321)^t$, where P is the U.S. population in millions and t is the number of decades since 1780. This model predicts $P = 9.3$ for 1820 and $P = 21.5$ for 1850.

   b. The prediction for 1820 is low by 0.3 million, and the prediction for 1850 is low by 1.7 million.

   c. For 1920 and 1950 the model predicts 151.2 million and 348.6 million.

   d. The predictions are high by 45.5 million and 197.9 million, respectively.

   e. The population has been growing more slowly than predicted by the model.

5. We can solve $3.055A^{0.310} = 25$ by computing $A = (25/3.055)^{(1/0.310)} = 880.91$, or about 881 square miles.

7. The model in Example 2 is $L = 2147.8(0.9909)^t$, where L is the L-dopa level in nanograms per milliliter and t is time in minutes.

   a. For $t = 200$ the model predicts $2147.8(0.9909)^{200} = 345$ nanograms per milliliter.

   b. Solving $100 = 2147.8(0.9909)^t$ for t gives $t = 335.5$ minutes.

9. **a.** Using a calculator, we find that the best-fit power function is $N = 1.506A^{0.3293}$, where A is area in square kilometers.

   b. Solving $15 = 1.506A^{0.3293}$, we find that an area of $A = 1075$ square kilometers is required to support 15 species of nonflying mammals.

11. **a.** The models are (with $t = 0$ in 1975):
     Linear: $C = 255t + 11$
     Exponential: $C = 15.29(1.0735)^t$
     Power: $C = 6.76t^{0.747}$
     b. The predictions for 2005 are 87.5, 128.37, and 84.90.

13. **a.** With $t = 0$ corresponding to 1950, the models for the number of violent crimes per 100,000 people are
     Linear: $N = 12.9203t - 272.9816$
     Exponential: $N = 133.392(1.0428)^t$
     b. The linear model predicts 848 crimes per 100,000 in 2000; the exponential model predicts 1084.
     c. The linear model predicts that the number of violent crimes will reach 1000 per 100,000 in 2009; the exponential model predicts this will happen in 1998.
     d. To find the doubling time, solve $2 = (1.0428)^t$, which gives $t = 16.5$.

15. **a.** The models for the number of college graduates in thousands as a function of years since 1890 are:
     Linear: $N = 12.9203t - 272.9816$
     Exponential: $N = 26.0174(1.0405)^t$
     Power: $N = 0.3347t^{1.7269}$
     Logarithmic: $N = -1455.3584 + 501.7958\ln t$
     b. The predictions for 2005 are: linear, 1213; exponential, 2498; power, 1211; logarithmic, 926.
     c. Linear and power both give reasonable predictions.
     d. The predicted years in which there will be 2 million college graduates are: linear, 2065; exponential, 1999; power, 2043; logarithmic, 2868.
     e. The doubling time is the solution of $2 = (1.0405)^t$, which is about 17.5 years.

17. **a.** See the table in the textbook.
     b. The models for the number of college degrees as a function of the number of high school diplomas are:
     Linear: $C = 0.335H - 40.3864$
     Exponential: $C = 52.7295(1.0011)^H$
     Power: $C = 0.2568H^{1.014}$
     Logarithmic: $C = -1402.8353 + 271.786\ln H$
     c. The slope for the linear fit indicates that the number of college degrees awarded increases by 0.335 for each additional high school degree awarded.
     d. The growth factor for the exponential model indicates that the number of college degrees grows at an increasing rate as the number of high school degrees grows.

19. The functions predict the cumulative HIV infections worldwide, in millions, with t representing the number of years since 1979.
     a. $N = 0.3072(1.32996)^t$
     b. The growth factor is 1.32996, indicating a growth rate of about 33% per year.
     c. Using the exponential model, the prediction is that 509.6 million people will have been infected by 2005.
     d. The function grows more slowly than the data from 1984 to 1995, then faster than the data.
     e. $N = 0.0657t^{2.1432}$
     f. 70.8 million
     g. The function matches well up to 1995, then grows more slowly than the data.
     h. The power function appears to be a better fit to the growth pattern.

21. For all three models, P is the percentage of the world's population living in urban areas and $t = 0$ corresponds to 1945.
    a. The best-fit linear model is $P = 0.3378t + 28.1705$.
    b. The year 2020 corresponds to $t = 75$ so the linear model predicts $0.3378(75) + 28.1705 = 53.5\%$ in urban areas.
    c. The best-fit exponential model is $P = 29.0134(1.0090)^t$.
    d. The exponential model predicts $29.0134(1.0090)^{75} = 56.8\%$ in urban areas.
    e. The best-fit power model is $P = 20.5929t^{0.1907}$.
    f. The power model predicts $20.5929\left(75^{0.1907}\right) = 46.9\%$ in urban area.
    g. The power model predicts a decline in urban residence, which is unlikely. The linear and exponential models both make reasonable predictions.

23. a. The best-fit logarithmic function is $T = 4.4258 + 19.528 \ln P$, where P is pressure and T is temperature.
    b. At a pressure of 6.2 kiloPascals the model predicts a boiling point of $4.4258 + 19.528 \ln(6.2) = 40.1°C$.
    c. Solving $120 = 4.4258 + 19.528 \ln P$ for P, we find $P = 371.8$ kiloPascals.

Section 3.4   How to Fit Exponential and Logarithmic Functions to Data

For each problem we indicate the data transformation required. Since the TI-83 calculator uses the transformation approach in it regression routines, we then give the best-fit function found by the calculator.

1. a. To develop an exponential fit, we do a linear fit of the logarithm of the population to the decade, as in Example 1. This converts to the exponential model $P = 3.003(1.3303)^t$, where t counts decades since 1780.
    b. The r-values for the original fit and the shorter-interval fit are 0.9982 and 0.9989, so the two fits are essentially equally good.

3. a. To develop an exponential fit, we do a linear fit of the logarithm of the debt to the number of years since 1940. The final model is $D = 216.97(1.0464)^t$.
    b. The year 2005 corresponds to $t = 65$, so the model predicts a debt of $216.97(1.0464)^{65} = 4138$ billion dollars.

5. The average debt in thousands of dollars is as follows:

Year	1940	1950	1960	1970	1980	1990	2000
Debt per person ($1000)	4.943	1.705	1.623	1.874	4.013	12.895	20.206

None of our standard model types gives a good fit, since the average debt per person first decreases and then increases. If we discard the data for 1940 the exponential model $A = 0.606(1.057)^t$ with $t = 0$ in 1940 gives a reasonable fit $(r = 0.936)$. For the year 2005 this model predicts an average debt per person of $0.606(1.057)^{65} = 22.251$, that is, $22,251.

7. If median family income grows at 5.5% and inflation continues at 3% we would expect the real growth to be about $\frac{1.055}{1.03} - 1 = 2.4\%$.

9. Answers will vary.

11. **a.** With $t = 0$ in 1980, the exponential model is $W = 52.4972(1.3730)^t$, where W is worldwide wind generating capacity in megawatts.

  b. The doubling time is the solution of $2 = (1.3730)^t$, which is about 2.2 years. The amount of wind energy produced doubles about every 2.2 years.

  c. According to the model the wind energy generating capacity in 2010 will be $52.4972(1.3730)^{30} = 708,357$ megawatts.

13. **a.** With $t = 0$ in 1960, the exponential model for atmospheric concentration of CO_2 in parts per million is $C = 314.254(1.0039)^t$.

  b. The doubling time is the solution of $2 = (1.0039)^t$, which is about 178 years. It will take 178 years for the carbon dioxide concentration to double.

  c. For 2010 the model predicts a concentration of $314.254(1.0039)^{50} = 381.8$ ppm.

  d. Solving $400 = 314.254(1.0039)^t$, we find $t = 61.9$, which corresponds to late in 2001.

15. **a.** With $t = 0$ in 1960, the exponential model for millions of telephones in use worldwide is $N = 91.1092(1.0602)^t$.

  b. The doubling time is the solution of $2 = (1.0602)^t$, which is about 11.9 years. The number of telephones is doubling about every 12 years.

  c. For the number of telephones in 2020 the model predicts $91.1092(1.0602)^{60} = 3040$ million.

  d. Solving $1000 = 91.1092(1.0602)^t$ gives $t \approx 41$, so the model predicts one billion phones by 2001.

  e. We will assume that the model $P = 3.6(1.013)^t$ takes 1970 as $t = 0$; the units are billions of people. We can adjust the telephone model to a starting year of 1970 by changing the constant multiplier to the number of phones in 1970, which according to the model is 163.47 million or 0.1635 billion. A telephone model with the same units and same starting year as the population model is then $T = 0.1635(1.0602)^t$. Solving $3.6(1.013)^t = 0.1635(1.0602)^t$ by taking logarithms, we find $t = 67.9$, which corresponds to the year 2038.

17. **a.** With $t = 0$ in 1985, the exponential model for the worldwide number of computers connected to the Internet, in thousands, is $N = 7.1195(2.0315)^t$.

  b. The doubling time is the solution of $2 = (2.0315)^t$, or 0.98. The number of connections is doubling about every year.

  c. For connections to the Internet in 2020, the model predicts $1000\left(7.1195(2.0315)^{35}\right) = 4.227 \times 10^{14}$; clearly the model is outside its useful range.

  d. Solving $250000 = 7.1195(2.0315)^t$, we get $t = 14.8$, so the model predicts 250 million connections by 2000.

Exercising Your Algebra Skills

For these problems we will assume that the transformations were carried out using base-10 logarithms. To re-transform to the original data from $Y = a + bX$ we write $y = 10^a \left(10^b\right)^x$.

1. $Y = 0.7782 + 0.0219X$ corresponds to $y = 10^{0.7782} \cdot (10^{0.0219})^x \Rightarrow y = 6.007(1.0517)^x$

2. $Y = 1.3010 + 0.0128X$ corresponds to $y = 19.9986(1.0299)^x$

3. $H = 1.0729 - 0.0223X$ corresponds to $y = 11.8277(0.9499)^x$

4. $Y = -0.3010 - 0.0706X$ corresponds to $y = 0.500(0.8500)^x$

5. $Y = 0.3522 + 1.0843X$ corresponds to $y = 2.2502(12.1423)^x$

6. $Y = -1.3015 + 0.7840X$ corresponds to $y = 0.0499(6.0814)^x$

7. $Y = 0.8525 - 1.2733X$ corresponds to $y = 7.1203(0.0533)^x$

8. $Y = -1.581 - 0.903X$ corresponds to $y = 0.0262(0.1250)^x$

Section 3.5   How to Fit Power Functions to Data

1. **a.** The model in Example 3 is $S = 14.35W^{0.3942}$, where W is takeoff weight in thousands of pounds and S is wingspan in feet. For a takeoff weight of 500 thousand pounds, the wingspan according to the model is $14.35\left(500^{0.3942}\right) = 166.3$ feet.

  b. For the super-jumbo jet, the required wingspan is $14.35\left(1200^{0.3942}\right) = 234.8$ feet.

  c. To find the maximum load for a wingspan of 175 feet, we solve $175 = 14.35W^{0.3942}$ by taking logarithms and find $\log 175 = \log 14.35 + 0.3942 \log W$. Then $\log W = 2.7554$ and $W = 10^{2.7554} = 569.4$ thousand pounds.

3. **a.** The best-fit power function is $D = 0.000385\, n^{1.4865}$, where n is the frequency on the dial and D is the distance from the left in centimeters.

  b. The model estimates the distance to 880 as $0.000385\left(880^{1.4865}\right) = 9.2$ cm and the distance to 1270 as $0.000385\left(1270^{1.4865}\right) = 15.8$ cm.

  c. To find the frequency 6 cm from the left of the dial, we solve $6 = 0.000385\, n^{1.4865}$ and obtain $n = 661.5$.

  d. Answers will vary.
  e. Answers will vary.

5. **a.** Answers will vary.

  b. The best-fit power function is $S = 18.4490\, L^{1.1526}$, where S is speed in cm/sec and L is length in cm.

  c. To find the length corresponding to a speed of 1500 cm/sec, we solve $1500 = 18.4490\, L^{1.1526}$ for L and obtain 45.4 cm.

  d. 6 feet is $2.54 \cdot 6 \cdot 12 = 182.9$ cm, so the model predicts a speed of $18.4490\left(182.9^{1.1526}\right) = 7471$ cm/sec, or about 75 m/sec. Since 10 seconds for the 100-meter dash is close to a world record, the model does not seem to apply to human runners.

7. **a.** The best-fit power model is $S = 10.2782\, L^{0.8600}$, where S is swimming speed in cm/sec and L is body length in cm.

  b. A fish that is 50 cm long should be able to swim at $10.2782\left(50^{0.8600}\right) = 297.2$ cm/sec.

  c. Solving $1000 = 10.2782\, L^{0.8600}$ for L, we find $L = 205.0$ cm.

9. a. The number of species depends on area, so we take area as the independent variable.

    b. The best-fit power model is $N = 63.7256\, A^{0.3521}$, where N is number of bird species and A is area in thousands of square miles.

    c. The model estimates $63.7256\left(4.6^{0.3521}\right) = 109$ species of birds.

    d. Solving $200 = 63.7256\, A^{0.3521}$ for A, we find a required area of 25.7 thousand square miles.

11. a. Angle depends on radius, so we take radius as the independent variable.

    b. A reasonable range would be from 0° to about 50°.

    c. The best-fit power model is $A = 31.4584\, R^{-0.8102}$, where A is the angle of lean in degrees and R is the radius in meters.

    d. Solving $15 = 31.4584\, R^{-0.8102}$ for R, we find that the corresponding radius is 2.5 meters.

    e. For a radius of 1.5 meters the required angle is $31.4584\left(1.5^{-0.8102}\right) = 22.6°$.

13. a. The best power fit for this data is $T = 0.0836\, R^{0.6216}$, where T is the time in seconds to identify, locate, and fire, and R is the range in meters.

    b. The model predicts a time of $0.0836\left(1750^{0.6216}\right) = 8.7$ seconds.

    c. Solving $10 = 0.0836\, R^{0.6216}$ for R, we find $R = 2201$ meters.

    d. According to the model, the crew has $0.0836\left(3000^{0.6216}\right) = 12.1$ seconds.

Exercising Your Algebra Skills

For these problems we will assume that the transformations were carried out using base-10 logarithms. To re-transform to the original data from $Y = a + bX$ we write $y = 10^a x^b$.

1. $Y = 0.90301 + 1.5X$ corresponds to $y = 10^{0.90301} x^{1.5} \Rightarrow y = 8.000\, x^{1.5}$

2. $Y = 0.6990 + 0.7X$ corresponds to $y = 5.000\, x^{0.7}$

3. $Y = 1.0792 - 1.5X$ corresponds to $y = 12.001\, x^{-1.5}$

4. $Y = -0.2218 - 0.4X$ corresponds to $y = 0.600\, x^{-0.4}$

5. $Y = 0.3522 + 1.0843X$ corresponds to $y = 2.250\, x^{1.0843}$

6. $Y = -1.3015 + 0.7840X$ corresponds to $y = 0.050\, x^{0.7840}$

7. $Y = 0.8525 - 1.2733X$ corresponds to $y = 7.104\, x^{-1.2733}$

8. $Y = -0.817 - 2.015X$ corresponds to $y = 0.152\, x^{-2.015}$

Section 3.6   How Good Is the Fit?

1. Since the linear fit removes a linear increasing or decreasing trend, the shape of the residuals depends only on the concavity of the data scatter. Where this is concave down, the linear fit will overestimate the data in the middle and underestimate it at the ends, so the residuals will be positive in the middle and negative at the ends. For data that is concave up, the pattern is reversed.

    a. (ii)   b. (i)   c. (ii)   d. (i)

3. a. To calculate residuals, subtract the predicted y-value from the data value. For example, for the point $(0, 1)$ the residual is $1 - (2 \cdot 0 + 1)$ or 0. The remaining residuals are 0, 1, 1, and 1. The sum of their squares is 3.

    b. For $y = 2.5x + 1$ the sum of squared residuals is 1.5, half the value found in part (a).

    c. The least-squares regression line is $y = 2.3x + 1$, with sum of squared residuals 0.3, much less than the sums of squares found in (a) and (b).

5. a. $S = 5602.99E - 43477.62$; each additional year of education corresponds to a salary increase of about $5600.

    b.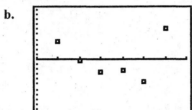

    The residual plot has a slightly concave up appearance, indicating concavity in the data which the linear fit does not capture.

    c. The sum of squared errors is 2.905×10^7.

7. a. The best power fit is $S = 111.5539\, E^{2.1533}$.

    b. The sum of squared errors for this model is 2.905×10^7.

    c.

    The residual plot shows a good fit for the first four points, less good for the last two.

9. a. Orbital speed is computed as the circumference of the planets circular path divided by the period of the planet's revolution, thus it is computed as $2\pi D/t$.

Planet	Period t	Distance D	Orbital Speed
Mercury	88	36.0	2.57
Venus	225	67.2	1.88
Earth	365	92.9	1.60
Mars	687	141.5	1.29
Jupiter	4329	483.3	0.70
Saturn	10753	886.2	0.52
Uranus	30660	1782.3	0.37
Neptune	60150	2792.6	0.29

    b. By examining the plots of the linear, exponential, and power function fits or by studying correlation coefficients, we see that the power function fit is superior. The formula the regression gives is $11.48t^{-0.334}$.

    c. Kepler's third Law states: $t^2 = 0.1664D^3$, or $D = 1.82t^{2/3}$. We calculated orbital speed as $2\pi D/t$. Substituting the value of t into this computation yields $6.28 \times 1.82t^{-1/3} = 11.42t^{-1/3}$. This formula differs from the power curve fit of part (b) only because of the accuracy of the original data.

Section 3.7   Linear Models with Several Variables

1. **a.** The coefficient of determination is 0.996, indicating that 99.6% of the variation is explained by the linear model; the multiple correlation coefficient is 0.998.
    b. $y = 6.284x_1 - 0.435x_2 + 34.561$
    c. For $x_1 = 25.5$ and $x_2 = 49.9$, the predicted height is $6.284(25.5) - 0.435(49.9) + 34.561 = 173.1$ cm vs. an actual height of 172 cm.

3. **a.** The coefficient of determination is 0.916, indicating that 91.6% of the variation is explained by the linear model; the multiple correlation coefficient is 0.957.
    b. $y = 0.7301x_1 + 0.547x_2 - 0.1642$
    c. The model predicts a forced expiratory volume of $y = 0.7301(2.6) + 0.547(3.9) - 0.1642 = 1.9$ liters per second.
    d. Vital lung capacity has a larger coefficient, and since the inputs for the two variables are about the same size, this suggests that lung capacity has a greater effect on forced expiratory volume.

5. **a.** The coefficient of determination is 0.388, indicating that 38.8% of the variation is explained by the linear model; the multiple correlation coefficient is 0.623.
    b. $y = 0.1620x_1 + 0.7117x_2 - 5.891$
    c. For an executive with 8 years in a company employing 484, this model predicts a stress level of $0.1620(484) + 0.7117(8) - 5.891 = 78.2$.
    d. Using the first three variables, $R^2 = 0.739$ and $R = 0.859$; the model now explains about 74% of the variation. The equation is now $y = 0.1882x_1 - 2.9607x_2 + 1.9798x_3 - 141.501$, and the predicted stress level if we know in addition that the manager's salary is $81,000 is 86.3.
    e. Using all four variables, $R^2 = 0.852$ and $R = 0.923$; the model now explains about 85% of the variation. The equation is now $y = 0.1702x_1 - 1.2035x_2 + 1.4919x_3 + 1.8644x_4 - 195.431$, and the predicted stress if we know in addition that the manager is 40 years old is 72.8.

Chapter 3 Review Problems

1. Let A = attendance and B = budget.
    The linear model is given by $A = 0.11101B + 0.25811$.
    The exponential model is given by $A = 0.50077(1.080835)^B$.
    The power function model is given by $A = 0.167755\,B^{0.875746}$.
    The logarithmic model is given by $A = -1.256558 + 2.829569 \log B$.

2. **a.** For all models we let $t = 0$ correspond to 1979.
    Linear: $H = 51.79t + 159.97$
    Exponential: $H = 251.92(1.089)^t$
    Power: $H = 218.35t^{0.532}$
    Logarithmic: $H = 124.91 + 300.89 \ln t$
    b. The respective predictions for 2004 ($t = 25$) are $1454.7 billion, $2123.0 billion, $1210.2 billion, and $1093.4 billion.
    c. Answers will vary; the log prediction is clearly too low.

3. **a.** For all models we let $t = 0$ correspond to 1979.

    Linear: $H = 13.90t + 320.47$

    Exponential: $H = 332.11(1.0311)^t$

    Power: $H = 322.05\,t^{0.1824}$

    Logarithmic: $H = 310.59 + 80.945\ln t$

    b. The respective predictions for 2004 ($t = 25$) are \$668.0 billion, \$714.2 billion, \$579.3 billion, and \$571.1 billion.

    c. Answers will vary.

4. **a.** The slopes indicate that health expenditures are increasing at about \$51.8 billion per year, and public education expenditures are increasing at \$13.9 billion per year.

    b. Solving $51.79t + 159.97 = 13.90t + 320.47$ we find $t = 4.2$, or the year 1983.

5. **a.** The bases indicate that health expenditures are increasing by 8.9% per year and education expenditures are increasing at 3.1% per year.

    b. Solving $251.92(1.089)^t = 332.11(1.0311)^t$ we find $t = 5$, or the year 1984. This is consistent with the information in the table, which indicates that by 1985 health expenditures exceeded education expenditures.

6. The best fit line for health expenditures as a function of education expenditures is $H = 3.68E - 1012.33$. Health expenditures are increasing by \$3.7 billion for each billion-dollar increase in education expenditures.

7. **a.** The best-fit power function for longevity in years as a function of gestation period in days is $L = 0.5432\,G^{0.6471}$, but as the scatter plot shows, humans represent a significant outlier. More useful fits might be obtained by discarding this data point.

    b.
    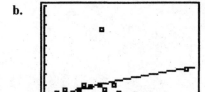

8. $P = 0.23w + 0.14$. This model changes continuously with weight, whereas the actual postage jumps by 23 cents as you go past each whole number of ounces.

9. **a.** The best-fit linear function is $C = 38.3991t + 45.8474$, where $t = 0$ in 1970.

    b. Predicted consumer credit at the end of 1992 is $38.3991(22) + 45.8474 = 890.6$ billion dollars.

    c. Solving $1500 = 38.3991t + 45.8474$ for t, we find $t = 37.9$, which corresponds to late in 2007.

10. **a.** The best-fit exponential is $N = 5.491(1.0298)^t$, with $t = 0$ in 1983.

    b. The fit has an r-value around 0.74, so it is not particularly good.

    c. Solving $10 = 5.491(1.0298)^t$, we find $t = 20.4$, so we predict 10 million flights in 2003 or 2004.

11. **a.** We would expect a linear fit to do well here: $H = 12.5877n + 20.8676$.

    b. The slope indicates the number of feet per story.

    c. A typical building in this list has about 12.5 feet per story.

12. The function *f* appears to be linear. The function *g* appears to be exponential. The function *h* appears to be logarithmic.

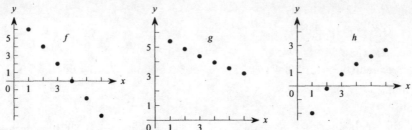

13. **a.** The best linear fit to the times *t* when the ball hits the floor as a function of the number of bounces *n* is $t = 1.03n + 0.0186$. The slope indicates that each complete bounce (floor to floor) takes about 1 second.

    b. $H = -1.0889n + 8.3482$. The slope here indicates that the bounce heights are decreasing about 1 foot for each bounce.

    c. The exponential $H = 9.0002(0.80007)^n$ provides a good fit. The base indicates that the height decreases 20% each time the ball bounces.

    d. The function $H = -1.007t + 8.695$ can be used to relate time to maximum bounce height.

    e. The next bounce will reach a maximum height of $9.0002(0.80007)^7 = 1.89$ feet at time $1.03(7.5) + 0.0186 = 7.74$ seconds. (We use a time halfway between the 7th and the 8th floor impacts as the time of the peak of bounce number 7.)

14. The model in Example 4 of Section 3.5 is $t = 0.4079 D^{1.5}$, where *D* is average distance from the sun in millions of miles and *t* is orbital period in days. For Quaoar, the model predicts a period of $0.4079\left(4000^{1.5}\right) = 103,191$ days, versus the observed value of 105,120 days.

15. Solving $14 = 2.317 W^{0.4442}$ for weight *W*, we get an estimate of 57.4 pounds.

Chapter 4 Extended Families of Functions

Section 4.1   Introduction to Polynomial Functions

1. **a.** 3   **b.** 4   **c.** 2   **d.** 8   **e.** 3   **f.** 6

3. **a.** Not a zero; $P(3) = 36$ **b.** Not a zero; $P(2) = 10$
    c. 1 is a zero.   **d.** 0 is a zero.
    e. Not a zero; $P(-1) = 6$ **f.** Not a zero; $P(-2) = 6$
    g. -3 is a zero.

5. Possible roots are $-3, -1, 2,$ and 4.

7. Using a function grapher, we find one zero at approximately 2.165.

9. There is one inflection point, and there are two turning points at approximately 0.59 and 1.41.

Exercising Your Algebra Skills

1. $(6x^3 - 5x^2 + 8) + (6x - 5x^2) = 6x^3 - 10x^2 + 6x + 8$

2. $(6x^3 - 5x^2 + 8) - (6x - 5x^2) = 6x^3 - 6x + 8$

3. $(5x^4 + 6x^3 + 7x - 11) + (-4x^3 - 9x^2 + 12) = 5x^4 + 2x^3 - 9x^2 + 7x + 1$

4. $(5x^4 + 6x^3 + 7x - 11) - (-4x^3 - 9x^2 + 12) = 5x^4 + 10x^3 + 9x^2 + 7x - 23$

5. $(10 - 4x + 5x^3 + 3x^4) + (5x^4 + 6x^3 + 7x - 11) = 8x^4 + 11x^3 + 3x - 1$

6. $(10 - 4x + 5x^3 + 3x^4) - (5x^4 + 6x^3 + 7x - 11) = -2x^4 - x^3 - 11x + 21$

7. $x(3x - 5) = 3x^2 - 5x$

8. $x(4x + 2) = 4x^2 + 2x$

9. $x(7 + 3x) = 3x^2 + 7x$

10. $x(6 - 5x) = -5x^2 + 6x$

11. $(x - 1)(x - 3) = x^2 - 4x + 3$

12. $(x - 2)(x - 5) = x^2 - 7x + 10$

13. $(x - 2)(x + 3) = x^2 + x - 6$

14. $(x + 4)(x + 3) = x^2 + 7x + 12$

15. $(x + 5)(x - 5) = x^2 - 25$

16. $(x - 3)(x + 3) = x^2 - 9$

17. $(x + 2)(x - 2) = x^2 - 4$

18. $(x-21)(x+21) = x^2 - 441$

19. $(x-1)^2 = x^2 - 2x + 1$

20. $(x-3)^2 = x^2 - 6x + 9$

21. $(x+2)^2 = x^2 + 4x + 4$

22. $(2x+5)^2 = 4x^2 + 20x + 25$

23. $(2x-6)^2 = 4x^2 - 24x + 36$

24. $(x+10)^2 = x^2 + 20x + 100$

Section 4.2   The Behavior of Polynomial Functions

1. The description suggests a graph like the one given below. Polynomials of degree n, where $n \geq 3$ with a positive leading coefficient would produce such a graph.

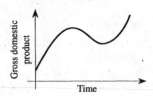

3. Since Graph **(i)** exits in opposite directions as x goes to positive and negative infinity, it is of odd degree. Since it has three roots, the minimum possible degree is 3. Since it goes to positive infinity as x goes to infinity, its leading coefficient is positive.

Since Graph **(ii)** exits in the same direction as x goes to positive and negative infinity, it is of even degree. Since it has four roots, the minimum possible degree is 4. Since it goes to positive infinity as x goes to infinity, its leading coefficient is positive.

Since Graph **(iii)** exits in the same direction as x goes to positive and negative infinity, it is of even degree. Since it has three turning points, the minimum possible degree is 4. Since it goes to negative infinity as x goes to positive infinity, it has a negative leading coefficient.

Since Graph **(iv)** exits in opposite directions as x goes to positive and negative infinity, it is of odd degree. Since it has five roots, the minimum possible degree is 5. Since it goes to negative infinity as x goes to positive infinity, it has a negative leading coefficient.

4. a. Plotting the points on a grapher shows that the full graph must have at least four turning points, so the

5. a. Graph **(iv)**, since both involve single roots at -3, 1, and 3.
    b. Graph **(vi)**, since both involve single roots at -1, -2, and 2.
    c. Graph **(v)**, since both involve a cubic polynomial with a single real root at 1.
    d. Graph **(i)**, since both involve single roots at -3, -1, 1, and 3.
    e. Graph **(ii)**, since both involve a triple root at the origin and a single root at 3.
    f. Graph **(iii)**, since both involve single roots at 2 and 4, and a double root at -3.

7. $f(x) = 5(x^2 - 4)(x^2 - 25)$

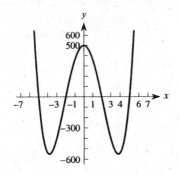

9. $f(x) = 5(x - 4)^2(x^2 - 25)$

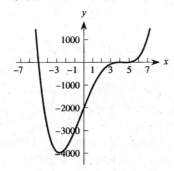

11. **a.** The polynomial P has roots at -3, 1, and 3. Thus it has linear factors $(x + 3)$, $(x - 1)$, and $(x - 3)$. Since the polynomial is required to be cubic, these are exactly the variable factors, and the polynomial may be written as $P(x) = a(x + 3)(x - 1)(x - 3)$, where a is some real number. To determine the value of a, we note that $P(0) = 2$, hence $2 = a(3)(-1)(-3)$. Solving for a, we get 2/9 and $P(x) = \frac{2}{9}(x + 3)(x - 1)(x - 3)$.

   b. The polynomial Q has a double root at $x = -1$ and a single root at $x = 3$. This polynomial may be written as $Q(x) = (x + 1)^2(x - 3)$. To find the value for a, we note that the y-intercept is 1. So, we get $Q(0) = a(0 + 1)^2(0 - 3) = 1$, which gives $a = -1/3$. Thus, the required polynomial is $Q(x) = -\frac{1}{3}(x + 1)^2(x - 3)$.

13. For each polynomial, a graph will indicate the number of real roots; the remaining roots will be complex.

   i. $P(x) = x^4 - 8x^2 - 9$ has two real roots, 3 and -3. Since the degree is 4, there are two other roots, which are complex.

   ii. $P(x) = x^5 - 4x^4 - 6x^3 + 6x^2 - 27x + 27$ has three real roots, at approximately $-2,386$, 0.929, and 5.108. Since the degree is 5, there are two other roots, which are complex.

   iii. $P(x) = x^6 - 4x^5 + 6x^4 - 16x^3 + 11x^2 - 12x + 6$ has two real roots, at approximately 0.586 and 3.414. Since the degree is 6, there are four other roots, which are complex.

   iv. $P(x) = x^6 - 9x^5 + 26x^4 - 41x^3 + 71x^2 - 42x + 6$ has four real roots, at approximately 0.209, 0.586, 3.414, and 4.791. Since the degree is 6, there are two other roots, which are complex.

15. **a.** The turning points are when $x \approx -0.5582$, $x \approx 0.76852$, and $x \approx 2.03967$.

   b. $P(x)$ is decreasing on $-\infty < x < -0.5582$ and on $0.76852 < x < 2.03967$. $P(x)$ is increasing on $-0.5582 < x < 0.76852$ and on $2.0367 < x < \infty$.

   c. There are two points of inflection located at $(0, -10)$ and $(1.5, -9.625)$.

   d. $P(x)$ is concave up on $-\infty < x < 0$ and on $1.5 < x < \infty$. $P(x)$ is concave down on $0 < x < 1.5$.

17. The parabola is necessarily pointing down, and its axis of symmetry passes through the point $(1, 5)$. Then the point $(0, 0)$ belongs to the graph of this quadratic function since it is the point symmetric to the x-intercept $(2, 0)$. The general form of a quadratic function is $y = ax^2 + bx + c$. In our case, if $x = 0$, we know that $y = 0$. Then $a(0^2) + b(0) + c = 0$, so $c = 0$. Substituting the values $x = 1$, $x = 2$ and the corresponding values of y, we obtain the equations $a + b = 5$ and $4a + 2b = 0$. From the first equation, we get $a = 5 - b$. When we substitute this value of a in the second equation, we get $4(5 - b) + 2b = 0$. Simplifying and collecting terms, we obtain $b = 10$. Therefore $a = 5 - 10 = -5$. The required equation is $y = -5x^2 + 10x$.

19. **a.** The quadratic must have the factors $(x - 6)$ and $(x + 2)$, so a possible formula for the quadratic is $y = (x - 6)(x + 2) = x^2 - 4x - 12$. Since any nonzero multiple of this is also a possibility, $y = a(x - 6)(x + 2) = ax^2 - 4ax - 12a$.

b. The turning point is halfway between the real roots, that is, at $x = 2$. If $x^2 - 4x - 12$, then the turning point is $(2, -16)$.

c, If the quadratic has a maximum value of 20, then the graph is pointing down and the value of y at the turning point is 20. So, $a(2^2) - 4a(2) - 12a = 20$, or $-16a = 20$ and $a = -5/4$. The required equation is $y = -\frac{5}{4}(x - 6)(x + 2)$, or $y = -\frac{5}{4}x^2 + 5x + 15$.

d. If the quadratic has a minimum value of -20, then the graph is pointing up and the value of y at the turning point is -20. So, $a(2^2) - 4a(2) - 12a = -20$, or $-16a = -20$ and $a = 5/4$. The required equation is $y = \frac{5}{4}(x - 6)(x + 2) = \frac{5}{4}x^2 - 5x - 15$.

21. **a.** A sensible meaning we can attribute to "start" in this problem is the time $t = 0$. Substituting this into the expression for s, we see that the object starts at height 0.

b. We need to find some reasonable meaning for the notion of "hits the ground." It is plausible to interpret the ground level as having height 0, so we can find the moment of impact by setting our expression for s equal to 0: $v_0 t - (1/2)gt^2 = t(v_0 - (1/2)gt) = 0$. Solving this equation we get two solutions: $t = 0$, indicating the moment at which the object was thrown, and $t = 2v_0/g$, the moment at which the object strikes the ground. Thus, the total flight of our object lasts $2v_0/g$ seconds.

c. The object reaches its maximum height at the midpoint of its flight, when $t = v_0/g$ seconds.

d. The maximum height that the object reaches is obtained by substituting $t = v_0/g$ into our expression for s. Doing so, gives $v_0(v_0/g) - \frac{1}{2}g(v_0/g)^2 = v_0^2/g - v_0^2/2g = v_0^2/2g$ cm.

23. $P(x) = x^3 - 5x^2 + 3x + 7$ has three roots, approximately at -0.87, 2.21, and 3.66. Using these approximations, $P(x)$ factors as $(x + 0.87)(x - 2.21)(x - 3.66)$.

25. **a.** The average rate of change of f from $x = -2$ to $x = 3$ is, by the definition given, $\Delta y / \Delta x = [f(3) - f(-2)]/(3 - (-2)) = (15 - 0)/5 = 3$.

b. The average rate of change for each successive pair of points in the table is computed below:

between $x = -2$ and $x = -1$: $\Delta y / \Delta x = (f(-1) - f(-2))/(-1 - (-2)) = (3 - 0)/1 = 3$
between $x = -1$ and $x = 0$: $\Delta y / \Delta x = (f(0) - f(-1))/(0 - (-1)) = (0 - 3)/1 = -3$
between $x = 0$ and $x = 1$: $\Delta y / \Delta x = (f(1) - f(0))/(1 - 0) = (-3 - 0)/1 = -3$
between $x = 1$ and $x = 2$: $\Delta y / \Delta x = (f(2) - f(1))/(2 - 1) = (0 - (-3))/1 = 3$
between $x = 2$ and $x = 3$: $\Delta y / \Delta x = (f(3) - f(2))/(3 - 2) = (15 - 0)/1 = 15$

The average value of all these slopes is $(3 + (-3) + (-3) + 3 + 15)/5 = 3$.

c. Extending the table to include the value for $x = 4$, the average rate of change of f from $x = -2$ to $x = 4$ is $[f(4) - f(-2)]/[4 - (-2)] = f(4)/6$. The average rate of change of f from $x = 3$ to $x = 4$, is $[f(4) - f(3)]/[4 - 3] = (f(4) - 15)/1 = f(4) - 15$. The average of all the slopes between successive pairs of points is $[3 + (-3) + (-3) + 3 + 15 + f(4) - 15]/6 = f(4)/6$. We observe that the same result holds. The average of all the average rates of change between successive points is equal to the average rate of change over the whole interval.

    d. Extending the table to include the value for $x = -3$, the average rate of change of f from $x = -3$ to $x = 4$ is $[f(4) - f(-3)]/[4 - (-3)] = [f(4) - f(-3)]/7$. The average rate of change of f from $x = -3$ to $x = -2$ is the difference quotient $[f(-2) - f(-3)]/[-2 - (-3)] = [f(-2) - f(-3)]/1 = f(2) - f(-3) = 0 - f(-3) = -f(-3)$. The average of all the slopes between successive pairs of points is

$$[-f(-3) + 3 + (-3) + (-3) + 3 + 15 + f(4)-15]/7 = [f(4)-f(-3)]/7,$$

the same result as before.

    e. Our tentative theorem might state: Let f be a function defined on an interval $[a, b]$, and let $\{x_0 = a, x_1, x_2, \ldots, x_n = b\}$ be a collection of uniformly spaced points in the interval $[a, b]$. The average of all the average rates of change between pairs of successive points is equal to the average rate of change over the whole interval.

27. Any polynomial of degree $n \leq 2$ can be written in the form $p(x) = ax^2 + bx + c$. We can use this representation to reduce parts (a) – (d) to the problem of solving a system of equations that are linear in the parameters a, b and c. Throughout, we will take the equation $p(x) = ax^2 + bx + c$ as the formula for p.

    a. To require that $p(0) = 1$ gives the equation $p(0) = a(0)^2 + b(0) + c = 1$, or $c = 1$. Hence, $p(x) = ax^2 + bx + 1$. Continuing from this observation, we look at the second and third conditions: $p(1) = 1$ and $p(2) = 1$. Using the formula $p(x) = ax^2 + bx + 1$, we get the equations $a(1)^2 + b(1) + 1 = 1$ and $a(2)^2 + b(2) + 1 = 1$. These simplify to give the system $a + b = 0$, and $4a + 2b = 0$. Solving this system of equations simultaneously gives $a = 0$ and $b = 0$ as its only solution. For our original polynomial p we now know that $a = 0$, that $b = 0$, and that $c = 1$. Thus, $p(x) = 1$ is the only polynomial of degree n, where $n \leq 2$, that satisfies all three conditions $p(0) = 1$, $p(1) = 1$, and $p(2) = 1$. Okay, we admit that 1 is not much of a polynomial, but it *is* of the required form $ax^2 + bx + c$.

    b. Again taking $p(x) = ax^2 + bx + c$, the three conditions $p(0) = 1$, $p(1) = 1$, and $p(2) = 2$, we get the three equations $a(0)^2 + b(0) + c = 1$, $a(1)^2 + b(1) + c = 1$, and $a(2)^2 + b(2) + c = 2$. So, the given conditions gives the system of equations $c = 1$, $a + b + c = 1$, and $4a + 2b + c = 2$. This system has the unique solution $a = 1/2$, $b = -1/2$, $c = 1$. So, $p(x) = (1/2)x^2 - (1/2)x + 1$.

    c. As we saw in part (a), the conditions $p(0) = 1$ and $p(1) = 1$ result in the equations $c = 1$ and $a + b = 0$. From the second equation we can conclude that $b = -a$, but nothing more. There simply is not enough information given to determine unique values for a and b. Thus, any polynomial that can be written in the form $ax^2 - ax + 1$ satisfies the conditions $p(0) = 1$ and $p(1) = 1$.

    d. In this exercise, the given condition $p(0) = p(1)$ gives the constraint $c = a + b + c$. From this we can conclude that $b = -a$, but nothing more. The polynomials of degree $n \leq 2$ that satisfy this constraint are those that can be written in the form $p(x) = ax^2 - ax + c$, where a and c can be any real numbers.

Exercising Your Algebra Skills

  1. $x^2 + 7x + 12 = (x+3)(x+4)$

  2. $x^2 - 4x - 5 = (x-5)(x+1)$

  3. $x^2 - 7x + 12 = (x-3)(x-4)$

  4. $x^2 - x - 12 = (x-4)(x+3)$

  5. $x^2 + x - 12 = (x+4)(x-3)$

  6. $x^2 - 4x + 4 = (x-2)(x-2) = (x-2)^2$

  7. $x^2 - 6x + 9 = (x-3)(x-3) = (x-3)^2$

8. $x^2 - 25 = (x+5)(x-5)$

9. $x^2 - 100 = (x+10)(x-10)$

10. $x^2 + 36$ cannot be factored.

11. $x^3 + x^2 - 20x = x(x^2 + x - 20) = x(x+5)(x-4)$

12. $x^3 - 4x^2 + 3x = x(x^2 - 4x + 3) = x(x-3)(x-1)$

13. $x^3 + 10x^2 + 25x = x(x^2 + 10x + 25) = x(x+5)^2$

14. $x^3 - 36x = x(x^2 - 36) = x(x+6)(x-6)$

Section 4.3   Modeling with Polynomial Functions

1. a. Using the model $y = -221.9t^3 + 9261.8t^2 - 62275.9t + 122988.9$, where y is the total number of reported AIDS cases in the United States and t is the number of years since 1980, we predict the total number of cases through 2000 by substituting the value $2000 - 1980 = 20$ for t. This gives a prediction of 806,991 cases.

    b. According to the online edition of the 2001 Statistical Abstract of the United States at http://www.census.gov/prod/2002pubs/01statab/health.pdf, the total reported AIDS cases through 2000 were 774,467.

    c. If the cubic pattern continued, we would expect to find the total cases reported in 2004 by substituting the value $2004 - 1980 = 24$ for t. The result is 895,619 cases. However, note that the cubic model cannot be accurate at this point: our function has a turning point around $t = 23.9$, but the total number of reported cases cannot decrease.

    d. Solving by trial and error, we find that 850,000 total cases would be expected by time $t = 21.17$, that is, early in 2001.

3. If measurements are estimated from the left inner edge of the Gateway Arch instead of at the middle, all we need only to shift the x measurements 300 feet. The height measurements don't change. The data are:

x	−25	0	50	100	150	200	300	400	450	500	550	600	625
H	0	100	330	500	570	610	630	610	570	500	330	100	0

The best quadratic function to fit the data is $H = -0.006408x^2 + 3.884716x + 122.3078$. Note that another way to go about this is to take the quadratic found in the text and substitute $x + 300$ for x. The best quartic function to fit the data is $y = -0.0000000327x^4 + 0.00003927x^3 - 0.020492x^2 + 5.226436x + 125.361634$, which can also be obtained by substituting $x + 300$ for x in the quartic obtained in the text.

5. Taking the grid to be in the scale 1:1 and the origin to be at the lower left of the innermost arch, the estimated measurements of the height are:

x	0	1	2	3	4	4.6
Height	0	10	11.5	11.2	8	0

The best quadratic function that fits these data is $y = -2.308516x^2 + 10.729066x + 0.35989$. If we are happy with the assumption that the inner arch on the left is a copy of the one of the right, setting the origin on the lower right corner of the left arch gives us the data:

x	0	−1	−2	−3	−4	−4.6
Height	0	10	11.5	11.2	8	0

The best quadratic function that fits these data is $y = -2.308516x^2 - 10.729066x + 0.35989$, which could have been obtained by substituting $-x$ for x in the formula for the right arch.

7. The quadratic equation $T = -0.00001873R^2 + 0.0622R + 323.54$ provides a reasonable good fit, where T is the torque and R is the rpm.

9. A good fit is provided by $P = -0.03316t^2 + 0.7830t + 25.19$, with P being the population of 18- to 20-year-olds in millions and t being years after 1970. For the year 2000, we take $t = 30$ and get a prediction of $P = 18.84$ million. The year 2005 corresponds to $t = 35$, which yields a prediction of $P = 11.98$ million. We would expect the first prediction to be better, since the time is closer to the years covered by our data.

11. Taking the point where the road passes the first support as the origin, the quadratic fit is $h = 0.0001047d^2 - 0.4466d + 496.99$, where h is the height of the cable above the road in feet and d is the distance from the chosen origin. We can see from the first and last columns of the table that the distance between the two supports is 4300 feet, so instead we might subtract 2150 from each distance value in the table and take the center of the bridge roadway as the origin. This will produce an equation without a linear term.

13. a. The best-fit quadratic is $A = -0.1437t^2 + 5.9982t + 266.0148$, where A is kilograms of grain produced worldwide per person and t is years since 1965.
    b. The model has a maximum at about (21, 328.6), that is 328.6 kg per person in 1986.
    c. The model predicts a production level of 244.9 kg per person in 2010.

15. a. The best-fit quadratic is $D = -0.000003644T^2 - 0.00005986T + 1.0004$, where D is density and T is temperature above $0°C$.

    b. This function predicts a density of 0.97835 at $70°C$.

    c. Solving $-0.000003644T^2 - 0.00005986T + 1.0004$ for T, we find $T = 43.2°C$.

17. The distance D that a body falling travels in feet in t seconds is given by $D = 16t^2$. To find how long it takes an object to fall 179 feet, we need only solve $179 = 16t^2$ for t. It took 3.34 seconds for Galileo's objects to fall.

Section 4.4   The Roots of Polynomial Equations: Real or Complex

1. Whenever c is positive, $x^2 + c$ has only two complex roots. Whenever c is negative $x^2 + c$ has two real roots. Since 50% of the numbers are negative, we would expect 50% of these quadratics to have two real roots. And, yes, we are ignoring 0, but 0 is only one number out of infinitely many, so it is not going to change the percentage.

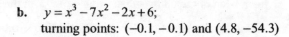

3.   The discriminant of $ax^2 + bx + c$ is $b^2 - 4ac$.

The discriminant of $(10a)x^2 + (10b)x + (10c)$ is $(10b)^2 - 4(10a)(10c) = 100(b^2 - 4ac.)$.

The discriminant of $(ka)x^2 + (kb)x + (kc)$ is $(kb)^2 - 4(ka)(kc) = k^2(b^2 - 4ac)$.

The roots of $(ka)x^2 + (kb)x + (kc)$ are:

$$\frac{-(kb) \pm \sqrt{k^2(b^2 - 4ac)}}{2ka} = \frac{-b \pm \sqrt{b^2 - 4ac}}{2a}.$$

So, the roots are the same as those of $ax^2 + bx + c$.

5.   **a.** $y = x^3 + 4x^2 - 8x + 3$;
      turning points: $(-3.4, 37.2)$ and $(0.8, -0.3)$

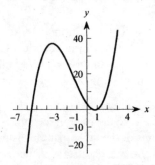

     b. $y = x^3 - 7x^2 - 2x + 6$;
      turning points: $(-0.1, -0.1)$ and $(4.8, -54.3)$

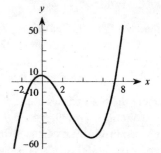

    c. $y = 5x^3 - 3x^2 - 6x + 8$;
      turning points: $(-0.4, 9.6)$ and $(0.9, 3.8)$

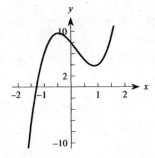

     d. $y = -4x^3 + 3x^2 + 5x - 4$;
      turning points: $(-0.4, -5.3)$ and $(0.9, 0.03)$

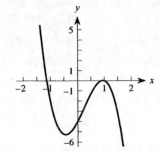

    e. $y = -4x^3 + 3x^2 - 5x - 4$;
      There are no turning points.

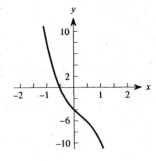

7. **a.** For $P(s) = s^2 + 6s + 8$, the roots are -4 and -2 (solve by factoring); the system is stable.

  b. For $P(s) = s^2 + 5s - 12$, by the quadratic formula, the roots are

$$\frac{-5+\sqrt{73}}{2} = 1.772 \text{ and } \frac{-5-\sqrt{73}}{2} = -6.772.$$

Since one root is positive and the other is negative, the system is unstable.

  c. For $P(s) = s^2 + 5s + 3$, the quadratic formula gives the two roots

$$\frac{-5+\sqrt{13}}{2} = -0.697 \text{ and } \frac{-5-\sqrt{13}}{2} = -4.303.$$

Since there are two real roots and both are negative, the system is stable.

  d. For $P(s) = s^3 - 4s^2 - 12s$, we see by inspection that $s = 0$ is a real root. Since this root is not negative, the system is unstable.

  e. For $P(s) = s^3 + 3s^2 + 7s + 5$, we try the factors of 5 as roots and find that -1 is a root. Factoring out $(s+1)$, we have $P(s) = (s+1)(s^2 + 2s + 5)$. Using the quadratic formula on $s^2 + 2s + 5 = 0$, we find that there are two complex roots, $-1+2i$ and $-1-2i$; since these both have a negative real part, this system is stable.

Section 4.5   Finding Polynomial Patterns

1. **a.**

x	y	Δy	$\Delta^2 y$
0	2		
1	0	-2	
2	4	4	6
3	15	11	7
4	30	15	4
5	52	22	7

Since the second differences are not constant, the given points do not lie on a parabola.

  b. Using the first three points, we would need to solve the system:

$$a \cdot 0^2 + b \cdot 0 + c = 2$$
$$a \cdot 1^2 + b \cdot 1 + c = 0$$
$$a \cdot 2^2 + b \cdot 2 + c = 4$$

The first equation tells that $c = 2$. Substituting this value into the second and third equations, we get

$$a + b = -2$$
$$4a + 2b = 2$$

This system has the solution $a = 3$ and $b = -5$. Thus, the first three points determine the quadratic equation $y = 3x^2 - 5x + 2$. This gives the last two points correctly, but for $x = 14$, $y = 14$ instead of 15 as required.

3. If these values are truly cubic, then the third differences will be constant. Computing the first differences we get Δy: $-6, -10, -2, 18, 50, 94, 160$. Computing the second differences gives $\Delta^2 y$: $-4, 8, 20, 32, 44$, and 66. The third differences are $\Delta^3 y$: $12, 12, 12, 12$, and 22. Clearly the last entry is at variance with the expected pattern. The problem is traceable to the last entry in the list of second differences. In order to get a last third difference of 12 without disturbing the other third differences the last entry needs to be 56, not 66. So, the corrected list of second difference is $-4, 8, 20, 32, 44, 56$. Well, this correction has consequences. There must have been an error in the list of *first* differences. Specifically, the final entry in the list of first differences must be off by 10. The corrected list of first differences is $-6, -10, -2, 18, 50, 94, 150$. This now allows us to identify the original error in the list of values of y. In order to produce 150 as the last first difference and not change any of the other entries, the last value of y needs to be 334.

5. Define $p(x) = ax^2 + bx + c$.. For $p(x)$ to pass through the points $(0, 1)$, $(1, 4)$ and $(2, 9)$ we must have

$$c = 1$$
$$a + b + c = 4$$
$$4a + 2b + c = 9$$

which has solution $a = 1$, $b = 2$, $c = 1$. At $x = 0.5$ and $x = 3$, the underlying function has approximate values $p(0.5) = 2.25$ and $p(3) = 16$, respectively.

7. **a.** $\displaystyle\sum_{k=1}^{25} k = \frac{25(25+1)}{2} = 325$ **b.** $\displaystyle\sum_{k=1}^{100} k = \frac{100(100+1)}{2} = 5050$

  c. $\displaystyle\sum_{k=1}^{1000} k = \frac{1000(1000+1)}{2} = 500,500$

9. Inasmuch as the grapefruit pyramid has a square base, the number of grapefruit in a pyramid that is n levels tall is $1 + 4 + 9 + \cdots + n^2 = (1/6)(n)(n+1)(2n+1)$. Evaluating (or graphing) this cubic we find that 13 layers will use 819 of the 1000 grapefruit. However if the store manager can find 15 more grapefruits, the manager can build a display with 14 amazing layers.

11. The relevant sum is

$$\sum_{t=0}^{90} R(t) = \sum_{t=0}^{90} -0.02t^2 + \sum_{t=0}^{90} 1.8t + \sum_{t=0}^{90} 42 = -0.02\sum_{t=0}^{90} t^2 + 1.8\sum_{t=0}^{90} t + 42\sum_{t=0}^{90} 1$$
$$= -0.02(247,065) + 1,8(4095) + 42(91)$$
$$= 6251.7 \text{ in.}$$

13. **a.** For the function $y = ax^2$, if x takes the uniformly spaced values $-2, -1, 0, 1, 2, 3$, then the corresponding values for y are $4a, a, 0, a, 4a, 9a$. The first differences are $\Delta y = -3a, -a, a, 3a, 5a$. The second differences are all constant: $\Delta^2 y = 2a$.
  b. For the function $y = ax^3$, if x takes the uniformly spaced values $-2, -1, 0, 1, 2, 3$, then the corresponding values for y are $-8a, -a, 0, a, 8a, 27a$. The first differences are $\Delta y = 7a, a, a, 7a, 19a$; the second differences are $\Delta^2 y = -6a, 0, 6a, 12a$; the third differences are all constant: $\Delta^3 y = 6a$.
  c. For the function $y = ax^4$, we obtain the fourth differences to be constant: $\Delta^4 y = 24a$.
  d. We predict that the fifth differences $\Delta^5 y$ for the function $y = ax^5$ are constant, with the constant value being $5! \cdot a = 120a$.

15. The difficulty in this problem is simply some fairly heroic bookkeeping. Since $y = ax^2 + bx + c$, we have $y_0 = ax_0^2 + bx_0 + c$, $y_1 = ax_1^2 + bx_1 + c$, and $y_2 = ax_2^2 + bx_2 + c$. Substituting these into the first formula of Problem 14 gives

$$\Delta^2 y_0 = y_2 - 2y_1 + y_0 = ax_2^2 + bx_2 + c - 2(ax_1^2 + bx_1 + c) + ax_0^2 + bx_0 + c.$$

Collecting like powers and factoring out the coefficients gives us the following pattern:

$$\Delta^2 y_0 = a(x_2^2 - 2x_1^2 + x_0^2) + b(x_2 - 2x_1 + x_0)$$

Okay, that got rid of c. Now we need to get Δx into this equation, so we are going to substitute $x_0 + \Delta x$ for x_1 and $x_0 + 2\Delta x$ for x_2.

$$\Delta^2 y_0 = a[(x_0 + 2\Delta x)^2 - 2(x_0 + \Delta x)^2 + x_0^2] + b[(x_0 + 2\Delta x) - 2(x_0 + \Delta x) + x_0]$$

Looking at the linear terms, we see that they are all going to cancel. That is enough to make us expand out the squares and see what happens:

$$\Delta^2 y_0 = a[(x_0 + 4x_0\Delta x + 4(\Delta x)^2) - 2(x_0 + 2x_0\Delta x + (\Delta x)^2) + x_0^2]$$

Now, all that is left is:

$$\Delta^2 y_0 = 2a(\Delta x)^2$$

Since all the second differences are constant and equal $\Delta^2 y_0 = \Delta^2 y$, making this substitution and solving for a gives

$$a = \frac{\Delta^2 y}{2(\Delta x)^2}.$$

17. Using our hard earned formula that we found in Problem 16, we have:

$$\sum_{k=1}^{25} k^3 = \frac{25^2 (25+1)^2}{4} = 105,625$$

19. **a.**

n	a_n	b_n	c_n		b_n	Δb_n	$\Delta^2 b_n$
1	3	4	5		4		
2	5	12	13		12	8	
3	7	24	25		24	12	4
4	9	40	41		40	16	4
5	11	60	61		60	20	4

Since the second differences $\Delta^2 b_n$ are constant, the b terms follow a quadratic pattern.

b. Solving the system

$$a(1)^2 + b(1) + c = 4$$
$$a(2)^2 + b(2) + c = 12$$
$$a(3)^2 + b(3) + c = 24$$

we find that the quadratic for b_n is $b_n = 2n^2 + 2n = 2n(n+1)$.

c. As we can see from the table, the formula for a_n is $a_n = 2n+1$. So, we need to show that $(a_n)^2 + (b_n)^2$ is the square of an integer, which will be c_n.

$$(2n+1)^2 + [2n(n+1)]^2 = 4n^2 + 4n + 1 + 4n^2(n^2 + 2n + 1)$$
$$= 4n^4 + 8n^3 + 8n^2 + 4n + 1$$

If this is the square of an integer, the first and last terms tell us that this integer should have the form $2n^2 + \cdots + 1$, and some trial and error shows that
$$4n^4 + 8n^3 + 8n^2 + 4n + 1 = (2n^2 + 2n + 1)^2.$$

So, $c_n = 2n^2 + 2n + 1$.

d. Using our formulas for a_n, b_n, and c_n, we find that the next Pythagorean triple in the table will be $(13, 84, 85)$.

Section 4.6   Building New Functions from Old: Operations on Functions

1. a. $f(5) + g(5) = 11 + 1/5 = 56/5$
    b. $f(5) - g(5) = 11 - 1/5 = 54/5$
    c. $f(5)g(5) = (11)(1/5) = 11/5$
    d. $f(5)/g(5) = 11/(1/5) = 55$
    e. $f(g(5)) = f(1/5) = -17/5$
    f. $g(f(5)) = g(11) = 1/11$
    g. $f(f(5)) = f(11) = 29$
    h. $g(g(5)) = g(1/5) = 5$
    i. $f(x) + g(x) = 3x - 4 + 1/x$
    j. $f(x) - g(x) = 3x - 4 - 1/x$
    k. $f(x)g(x) = (3x - 4)(1/x) = 3 - 4/x$
    l. $f(x)/g(x) = (3x - 4)/(1/x) = 3x^2 - 4x$
    m. $f(g(x)) = f(1/x) = 3/x - 4$
    n. $g(f(x)) = g(3x - 4) = 1/(3x - 4)$
    o. $f(f(x)) = f(3x - 4) = 3(3x - 4) - 4 = 9x - 16$
    p. $g(g(x)) = g(1/x) = 1/(1/x) = x$

3. a. $f(5) + g(5) = 10^5 + \log 5 = 100,000.699$
    b. $f(5) - g(5) = 10^5 - \log 5 = 99,999.30103$
    c. $f(5)g(5) = (10^5)\log 5 = 69,897.0004$
    d. $f(5)/g(5) = (10^5)/\log 5 = 143,067.6558$
    e. $f(g(5)) = f(\log 5) = 10^{\log 5} = 5$
    f. $g(f(5)) = g(10^5) = \log 10^5 = 5$
    g. $f(f(5)) = f(10^5) = 10^{10^5} = 10^{100,000}$
    h. $g(g(5)) = g(\log 5) = \log(\log 5) = -0.15554$
    i. $f(x) + g(x) = 10^x + \log x$
    j. $f(x) - g(x) = 10^x - \log x$
    k. $f(x)g(x) = 10^x(\log x)$
    l. $f(x)/g(x) = 10^x/(\log x)$
    m. $f(g(x)) = f(\log x) = 10^{\log x} = x$
    n. $g(f(x)) = g(10^x) = \log 10^x = x$
    o. $f(f(x)) = f(10^x) = 10^{10^x}$
    p. $g(g(x)) = g(\log x) = \log(\log x)$

5. Given: $f(2) = 10$ $f(4) = 20$ $f(6) = 35$ $g(2) = 8$ $g(4) = 4$ $g(6) = 2$
    a. $f(6) - f(4) = 35 - 20 = 13$; incorrect
    b. $f(g(6)) = f(2) = 10$; incorrect
    c. $g(g(6)) = g(2) = 8$; correct
    d. $f(2) - g(6) = 10 - 2 = 8$; correct
    e. $f(4) - g(4) = 20 - 4 = 16$; incorrect
    f. $f(4) \cdot g(4) = 20 \cdot 4 = 80$; incorrect
    g. $\dfrac{f(4)}{g(4)} = \dfrac{20}{4} = 5$; correct

7. a.

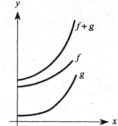

b.

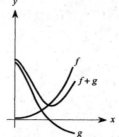

9. a. i.

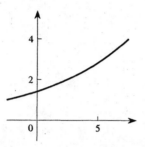

ii.

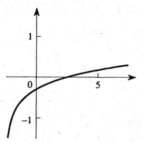

iii.

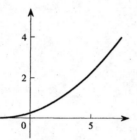

b. i.

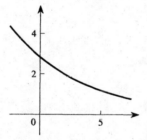

ii.

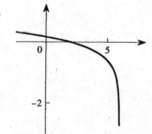

iii.

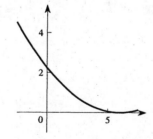

c.   **i.**

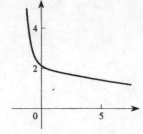

ii.

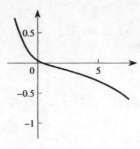

iii.

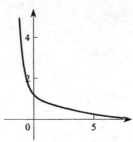

11.   **a.**

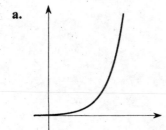

b.

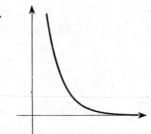

c.

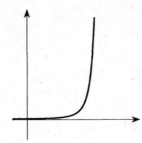

13.   **a.** $f(g(1)) = f(0) = 2$ **b.** $g(f(1)) = g(1) = 0$
      c. $f(g(-1)) = f(3) = -2$ **d.** $g(f(-1)) = g(1) = 0$

15. $G(x) = x^4$ and $F(x) = x + 5$ is one such pair.

17. $G(x) = x + 3$ and $F(x) = \log x$ is one such pair.

19.   We use the formula $t = 1 + \dfrac{s}{20} + \dfrac{70}{s}$.

   a. For 30 mph $= 44$ ft/sec, when $s = 44$, we have $t = 4.8$ seconds.

   b. For 50 mph $= \frac{5}{3} \cdot 44 = 73.33$ ft/sec, when $s = 73.33$, we have $t = 5.6$ seconds.

   c. A reasonable range would be 15 mph to 55 mph, which is the same as s being between 22 ft/sec and 81 ft/sec. The range for t might be from 0 sec to 7 sec, though when using this formula, the t values get no lower than about 4.7 sec.

d. It could decrease 20 or increase 70, or both.

e. A common denominator with be $20s$: $t = \dfrac{s^2 + 20s + 1400}{20s}$.

21.

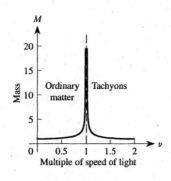

23. **a, b.** No, they are not the same. The point of these exercises is to emphasize that a square root does *not* distribute over a sum.

c. $\sqrt{x^2 + a^2} = |x| + a$, which in turn is equal to $x + a$ for nonnegative values of x only in the trivial case $a = 0$.

25. For $f(x) = \dfrac{x+1}{x}$:

a. $f(1) = 2$ **b.** $f(f(1)) = \frac{3}{2}$ **c.** $f(f(f(1))) = \frac{5}{3}$

d. The next two iterates are $\frac{8}{5}$ and $\frac{13}{8}$. The denominator of each fraction is the preceding numerator, and the numerator of each fraction is the sum of the preceding numerator and denominator. (You may recognize that both numerators and denominators follow the Fibonacci sequence.)

e. The fractions appear to be approaching a number close to 1.62. In fact the limit is the golden ratio,
$$\frac{1 + \sqrt{5}}{2} = 1.618034\ldots \, .$$

27. **a.** $V = \dfrac{1}{6\sqrt{\pi}} S^{3/2}$. One natural way of writing this as a composite function is $V(r(S))$, where
$$V(r) = \tfrac{4}{3}\pi r^3 \text{ and } r(S) = \frac{\sqrt{S}}{2\sqrt{\pi}}.$$

b. Solving the formula in part (a) for S yields $S = 6^{2/3}\pi^{1/3}V^{2/3}$. Among other things, this tells you that the surface area of a sphere is proportional to the two-thirds power of the volume.

29. Tracing through the given tables of function values, we find:
a. $g(f(A)) = G$
b. $f(g(A)) = L$
c. $f(f(P)) = Q$
d. $g(g(K)) = X$
e. We start in the g table looking for A on the right-hand side. $G(Q) = A$, so $g^{-1}(A) = Q$. Now we look for Q on the right-hand side in the f table and find $f(L) = Q$. Thus, $f^{-1}(Q) = f^{-1}\left[g^{-1}(A)\right] = L$.

Section 4.7   Building New Functions from Old: Shifting, Stretching, and Shrinking

1.   The completed table is

x	$f(x)$	$5f(x)$	$f(x)+3$	$f(x-1)$	$(f(x))^2$
3	5	25	8	undefined	25
4	2	10	5	5	4
5	−1	−5	2	2	1
6	3	15	6	−1	9
7	8	40	11	3	64

3.   **a.** $2g(x)$

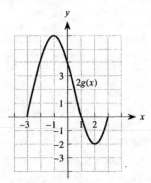

b. $f(x)+1$

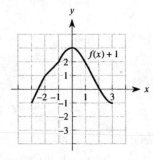

c. $f(x+1)$

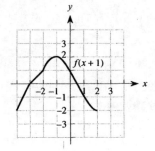

d. $f(x-1)$

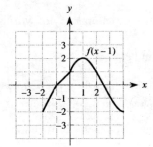

e. $g(2x)$

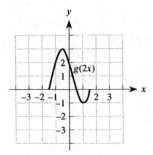

f. $g(x/2)$

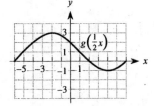

5.   **a.** The translation of the line $y = mx$ to a line with the same slope that passes through the point $(5, 12)$ is
$y = m(x - 5) + 12$.

     b. The translation of the line $y = mx$ to a line with the same slope that passes through the point (x_0, y_0) is
$y = m(x - x_0) + y_0$. This is known as the *point–slope* form of the equation of a line.

7.   **a.**

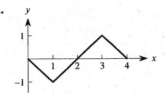

                                            b.

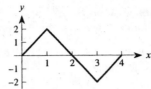

   c.

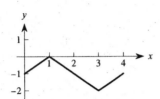

                                            d.

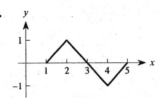

   e.

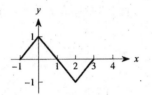

                                            f.

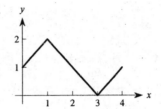

9.   The original table is:

x	−1	0	1	2	3	4	5	6	7
y	24	16	11	8	9	15	27	39	35

Shifting right 4 units add 4 to all the x values, and shifting up 7 units adds 7 to all the y values. The new
table is:

x	3	4	5	6	7	8	9	10	11
y	31	23	18	15	16	22	34	46	42

11.   **a.** $f(x) + h = x^2 - 3x + 4 + h$
      b. $f(x + h) = (x + h)^2 - 3(x + h) + 4 = x^2 + 2xh + h^2 - 3x - 3h + 4$
      c. $f(x + h) - f(x) = 2xh + h^2 - 3h = (2x + h - 3)h$
      d. $[f(x + h) - f(x)]/h = 2x + h - 3$
      e. Setting $x = 2$ and $h = 0.1$ in part (d) yields 1.1. Setting $x = 2$ and $h = 0.01$ in part (d) yields 1.01. Setting
$x = 2$ and $h = 0.0001$ in part (d) yields 1.0001.

13.   We assume the algebraic form of the function that describes the temperature of the turkey to be
$T = 350 - Ba^t$. Since at the time $t = 0$ the temperature is $T = 40°$, we get $B = 310$. Now, after one hour, or
$t = 60$, we have $T = 124°$. Therefore, $124 = 350 - 310a^{60}$. That is, $a^{60} = (350 - 124)/310$, so
$a \approx 0.99474656$. Thus, $T \approx 350 - 310(0.9947)^t$.

15. The information about the turning points is given in the following table:

n	1	2	3	4	5
turning point	1.4427	2.8854	4.3281	5.7708	7.2135

A linear regression equation is $y = 1.4427n$, so the predicted turning point for $f(x) = x^{1.5}(0.5)^x$ is $1.4427(1.5) = 2.16405$, which is accurate compared to the actual graph value of 2.1640408.

17. If actual data about the infection levels are available, an analysis of residuals can be carried out for several choices of these functions.

Exercising Your Algebra Skills

1. $f(2x) = (2x)^2 - 5(2x) + 3 = 4x^2 - 10x + 3$

2. $f(3x) = (3x)^2 - 5(3x) + 3 = 9x^2 - 15x + 3$

3. $f(4x) = (4x)^2 - 5(4x) + 3 = 16x^2 - 20x + 3$

4. $f(\frac{1}{2}x) = (\frac{1}{2}x)^2 - 5(\frac{1}{2}x) + 3 = \frac{1}{4}x^2 - \frac{5}{2}x + 3$

5. $f(x+1) = (x+1)^2 - 5(x+1) + 3 = x^2 + 2x + 1 - 5x - 5 + 3 = x^2 - 3x - 1$

6. $f(x-2) = (x-2)^2 - 5(x-2) + 3 = x^2 - 4x + 4 - 5x + 10 + 3 = x^2 - 9x + 17$

7. $f(2x-1) = (2x-1)^2 - 5(2x-1) + 3 = 4x^2 - 4x + 1 - 10x + 5 + 3 = 4x^2 - 14x + 9$

8. $f(x^2) = (x^2)^2 - 5(x^2) + 3 = x^4 - 5x^2 + 3$

Section 4.8   Using Shifting and Stretching to Analyze Data

1. Repeating the fit with 0.05 subtracted from the T values gives the equation $T = 8.65 + 45.2015(0.8292)^t$, which produces a slightly worse fit to the shifted data.

3. The best exponential fit is $T = 193.82565(0.96937)^t$ with $r = -0.991781$. The it is at best problematical since it predicts that the coffee freezes (at $32°$) in a room at $70°$.

5. **a.**

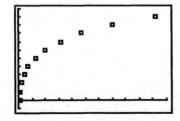

We can transform the data by subtracting the temperature values from 110 and fitting the resulting data with an exponential. We get $110 - T = 88.69(0.978)^P$, which can be rearranged to give an equation for T: $T = 110 - 89.69(0.978)^P$. Note that while this produces the desired asymptote, the fit is poor near the origin.

b. The boiling point corresponding to a pressure of 6.2 kiloPascals is $110 - 89.69(0.978)^{6.2} = 31.87$, or $31.87°C$.

   c. We need to solve $110-89.69(0.978)^P = 105$, which is equivalent to $\dfrac{5}{89.69} = (0.978)^P$.

      Taking logs on both sides and dividing, we have

$$P = \frac{\log(5/89.69)}{\log(0.978)} = 129.77$$

      or a pressure of 129.77 kiloPascals.

Section 4.9   The Logistic and Surge Functions

  1.   **a.** From the table we would conclude that typical full-grown heights are 176.8 cm for males and 163.7 cm
       for females.

     b. A logistic fit for the male data is:

$$H = \frac{192.91}{1+1.876e^{-0.166t}}$$

       A logistic fit for the female data is:

$$H = \frac{171.52}{1+1.73e^{-0.211t}}$$

     c. Based on these logistic functions, full growth for men is 192.91 cm and full growth for women is
       171.52 cm.

  3.   **a.** Turning point is farther to the right and higher; function eventually decays at the same rate.
     b. Turning point is farther to the left and lower; function eventually decays at the same rate.
     c. Turning point is farther to the left and lower; function eventually decays at a faster rate.
     d. Turning point is farther to the right and higher; function eventually decays at a slower rate.

Chapter 4 Review Problems

  1. $f(x) = (x+3)(x-2)(x-4)$ **2.** $g(x) = (2-x)(x+3)(x+1)$

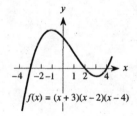
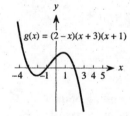

  3. $F(x) = (x+2)(x-3)(x-4)(x-1)$ **4.** $G(x) = (x+3)(x-2)(x-4)^2$

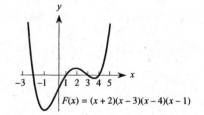
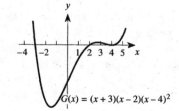

  5. $x^2 + x - 6 = (x+3)(x-2)$, so the zeros of this quadratic are $x = -3$ and $x = 2$.

  6. $2x^2 + 9x - 5 = (2x-1)(x+5)$, so the zeros of this quadratic are $x = 1/2$ and $x = -5$.

7. $x^3 - 3x^2 + 2x = x(x-1)(x-2)$; the three zeros of this cubic are $x = 0$, $x = 1$, and $x = 2$.

8. From Problem 5, $x^2 + x - 6$ has roots

    $$\frac{-1+\sqrt{1^2 - 4(1)(-6)}}{2(1)} = 2 \quad \text{and} \quad \frac{-1-\sqrt{1^2 - 4(1)(-6)}}{2(1)} = -3.$$

    From Problem 6, $2x^2 + 9x - 5$ has roots

    $$\frac{-9+\sqrt{9^2 - 4(2)(-5)}}{2(2)} = 0.5 \quad \text{and} \quad \frac{-9-\sqrt{9^2 - 4(2)(-5)}}{2(2)} = -5.$$

    From Problem 7, $x^3 - 3x^2 + 2x$ factors as $x(x^2 - 3x + 2)$, and so has one root, $x = 0$. The other roots are given by the quadratic formula:

    $$\frac{-(-3)+\sqrt{(-3)^2 - 4(1)(2)}}{2(1)} = 2 \quad \text{and} \quad \frac{-(-3)-\sqrt{(-3)^2 - 4(1)(2)}}{2(1)} = 1.$$

9. The vertex is at (15, 19). Since a parabola is symmetric around a vertical line through the vertex, the point (8, 4) is matched by the point (2, 4). Thus, $f(2) = 4$.

10. The inflection point is at $(6, -4)$. Since a cubic is symmetric around its inflection point, the point (2, 12) on the curve [right 4, up 16] is matched by the point $(2, -20)$ [left 4, down 16]. Thus, $f(10) = -20$.

11. The turning points of the graph of $y = x^3 + 4x^2 - 5$ are $(0, -5)$ and $(-8/3, 121/27)$.

12. **a.**
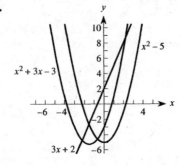
b.
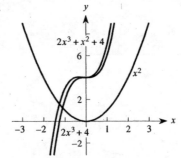

13. **a.** Zeros at $x = -2$, $x = 2$; turning point at $x = 0$; no vertical asymptotes; end behavior as x approaches ∞ is 1 and end behavior as x approaches $-\infty$ is 1.
    b. Zeros at $x = -2$, $x = 2$; turning point at $x = 3$; end behavior as x approaches ∞ is 1 and end behavior as x approaches $-\infty$ is 1.
    c. No zeros; turning point at $x = 0$; no vertical asymptotes; end behavior as x approaches ∞ is 1 and end behavior as x approaches $-\infty$ is 1.
    d. No zeros; turning point at $x = 0$; vertical asymptotes $x = -3$, $x = 3$; end behavior as x approaches ∞ is 1 and end behavior as x approaches $-\infty$ is 1.

14. **a.**

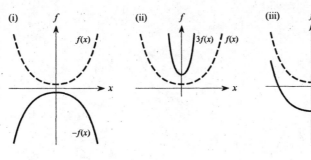

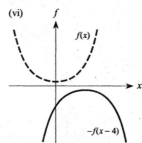

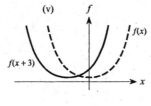

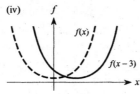

(i) $-f(x)$

(ii) $3f(x)$, $f(x)$

(iii) $f(x)$, $f(x) - 4$

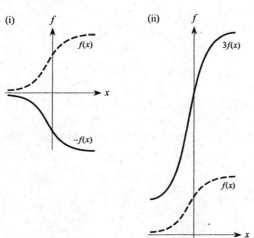

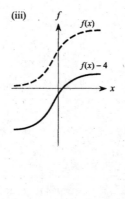

(iv) $f(x)$, $f(x - 3)$

(v) $f(x + 3)$, $f(x)$

(vi) $f(x)$, $-f(x - 4)$

b.

(i) $f(x)$, $-f(x)$

(ii) $3f(x)$, $f(x)$

(iii) $f(x)$, $f(x) - 4$

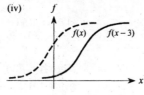

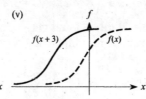

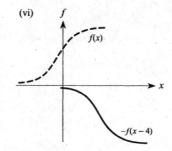

(iv) $f(x)$, $f(x - 3)$

(v) $f(x + 3)$, $f(x)$

(vi) $f(x)$, $-f(x - 4)$

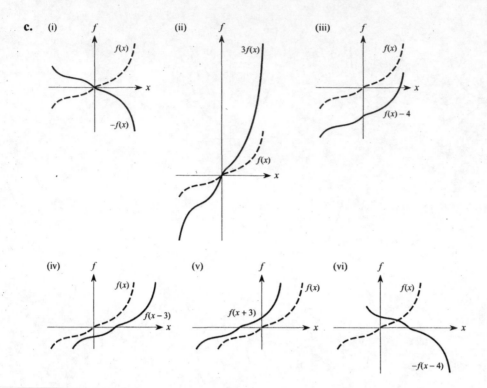

15. If $f(x) = 2x^2 + 1$ and $g(x) = (x-1)/(x+2)$, then:
    a. $f(3) + g(3) = 19.4$
    b. $f(f(3)) = 723$
    c. $g(f(3)) = 6/7$
    d. $g(g(3)) = -1/4$
    e. $g(3)f(3) = 7.6$
    f. $f(3)/g(3) = 47.5$
    g. $f(g(x)) = 3(x^2 + 2)/(x + 2)^2$
    h. $f(f(x)) = 8x^4 + 8x^2 + 3$
    i. $g(f(x)) = 2x^2/(2x^2 + 3)$
    j. $g(g(x)) = -1/(x + 1)$
    k. $g(x)f(x) = (x - 1)(2x^2 + 1)/(x + 2)$
    l. $f(x)/g(x) = (x + 2)(2x^2 + 1)/(x - 1)$

16. Given $f(0) = 2$, $f(1) = 2$, $f(2) = 3$, $f(3) = 0$ and $g(0) = 1$, $g(1) = 0$, $g(2) = 2$, $g(3) = 3$:
    a. $f(g(0)) = f(1) = 2$ $f(g(1)) = f(0) = 2$ $f(g(2)) = f(2) = 3$ $f(g(3)) = f(3) = 0$
    b. $g(f(0)) = g(2) = 2$ $g(f(1)) = g(2) = 2$ $g(f(2)) = g(3) = 3$ $g(f(3)) = g(0) = 1$
    c. $f(0) + g(0) = 3$ $f(1) + g(1) = 2$ $f(2) + g(2) = 5$ $f(3) + g(3) = 3$
    d. $f(0)/g(0) = 2$ $f(1)/g(1)$ is undefined $f(2)/g(2) = 3/2$ $f(3)/g(3) = 0$

17. From the graphs we get $f(1) = 1$, $f(2) = 3$, $f(3) = 3$, $f(4) = 1$ and $g(1) = 3$, $g(2) = 2$, $g(3) = 1$, $g(4) = 1$. We don't have definitions for $f(0)$ and $g(0)$. We evaluate the combinations of functions indicated in Problem 16 where possible.
    a. $f(g(1)) = f(3) = 3$ $f(g(2)) = f(2) = 3$ $f(g(3)) = f(1) = 1$ $f(g(4)) = f(1) = 1$
    b. $g(f(1)) = g(1) = 3$ $g(f(2)) = g(3) = 1$ $g(f(3)) = g(3) = 1$ $g(f(4)) = g(1) = 3$
    c. $f(1) + g(1) = 4$ $f(2) + g(2) = 5$ $f(3) + g(3) = 4$ $f(4) + g(4) = 2$
    d. $f(1)/g(1) = 1/3$ $f(2)/g(2) = 3/2$ $f(3)/g(3) = 3$ $f(4)/g(4) = 1$

18. For $0.1063 < t < 3.294$, approximately, the return of the second investment, $G(t) = 7.8t + 3.5$ is higher than the return of the first investment, $F(t) = 2t^2 + t + 4.2$.

19. The sum $3 + 6 + 9 + 12 + 15 + \cdots + 300$ can be expressed as $3(1 + 2 + 3 + 4 + 5 + \cdots + 100)$, and we have a formula for the sum of the first 100 positive integers, namely $100(100 + 1)/2$. Then, the given sum is $3[100(100 + 1)/2] = 15{,}150$.

20 a. It must have 3 inflection points.
    b. The minimum degree is 5.

c. Since the function either heads for plus infinity as the variable increases and minus infinity as the variable decreases, or vice versa, the graph must cross the x-axis at least one, so the minimum number of real roots is 1.

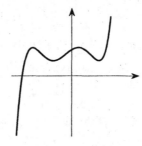

d. It could happen that the graph crosses the x-axis between each pair of turning points, to the left of the leftmost turning point, and to the right of the rightmost turning point. In this case, the polynomial would have 5 real roots, which is the maximum number.

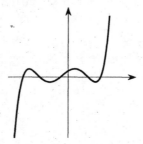

e. 2, 3, and 4 are also possible numbers of real roots, as indicated in the graphs below. Note that when the number of roots is even, one or more of the turning points must lie on the x-axis.

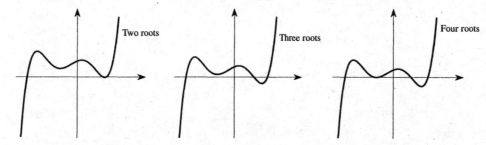

Two roots   Three roots   Four roots

21. To use quartic regression, we will need to estimate coordinates for at least five points on the graph. Using the points $(-3, 20)$, $(-2, 14)$, $(0, 4)$, $(1, 8)$ and $(2, 10)$, we find the equation $y = -x^4 + 6x^2 - x + 4$.

22. a.

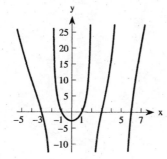

b. The three points where R is undefined are zeros of the denominator, so the denominator might be $(x+2)(x-2)(x-5)$. The five points where R is 0 are zeros of the numerator, so the numerator might be $(x+3)(x+1)(x-1)(x-3)(x-6)$. Experimenting shows that the function

$$R(x) = \frac{(x+3)(x+1)(x-1)(x-3)(x-6)}{(x+2)(x-2)(x-5)}$$

fits the remaining points in the table quite closely, though not exactly.

23. **a.** Graphs (i) and (iv) are exponential; Graphs (ii), (v), and (vi) are power; Graph (iii) is logarithmic.
   b. There are many possible answers. For example,

   (i) $f(x) = 2^x - 11$ (ii) $f(x) = \sqrt{x} + 5$ (iii) $f(x) = \log(x+4)$
   (iv) $f(x) = \left(\frac{1}{3}\right)^x + 7$ (v) $f(x) = x^{-1} + 9,\ x > 0$ (vi) $f(x) = (x-5)^{-3},\ x > 5$

24. **a.** $N = 4.8796(1.3480)^t$, where $t = 0$ in 1990.
   b. $B = 82.7650(0.9170)^t$, where $t = 0$ in 1990.
   c. $R = 398.8942(1.2691)^t$, where $t = 0$ in 1990.
   d. The new last row showing total annual revenue based on the formula in part (c) is:

   398.89   506.24   815.35   1034.76   1313.22   1666.61   2115.09   2684.26   3406.59

   The equation for the exponential fit to this data is $R = 398.8901(1.2691)^t$, where $t = 0$ in 1990. This is essentially the same as the function in part (c).

Chapter 5 Modeling with Difference Equations

Section 5.1 Eliminating Drugs from the Body

1. If 30% of a drug is removed every four hours, and the patient takes a single dose of 16 mL, then the level of the drug after 12 hours is given by $16(0.7)^3 \approx 5.5$ mL since 12 hours corresponds to 3 removal periods. To determine the level of the drug after 24 hours, we need to compute $16(0.7)^6 \approx 1.9$ mL. For the level of the drug to drop below 1 mL, we need the number of periods n of four hours such that $16(0.7)^n < 1$, which takes $n = 7.77$ periods of four hours, or approximately 31.1 hours. Similarly, we determine that it takes about 82.7 hours for the level of the drug to drop below 0.01 mL. This exercise corresponds to a simple exponential decay model since no extra dose is administered after the initial one.

3. a. $D_0 = 320$ and $D_1 = 0.75D_0 + 80 = 320$. The calculations for every subsequent period will look the same, so $D_2 = D_3 = D_4 = D_5 = 320$.

    b. $(0.75)(320) = 240 = 320 - 80$, so the daily dose of 80 mg exactly replaces the 80 mg eliminated each day.

5. The graph below shows dosage over continuous time intervals for 13 periods. The drug level decreases exponentially between the periodic dosages.

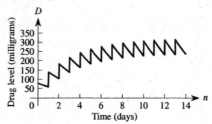

7. Here, $B = 10$ and $a = 1 - 0.60 = 0.4$. The maintenance level is $L = \dfrac{B}{1-a} = \dfrac{10}{0.6} = 16.67$ mg.

9. Taking 10 mg every two days does not achieve the same maintenance level as taking 5 mg every day. For a one-day dosage cycle, the percentage not eliminated is 40%. For a two-day cycle, we have 40% of 40% remaining, that is, $a = (0.40)^2 = 0.16$. With a dosage of 10 mg, the limiting value becomes $L = \dfrac{10}{1-0.16} = 11.9$ mg.

11. In Problem 10 we found that the amount of aspirin in the bloodstream decreases by a factor of 0.2383 per hour. After 24 hours there will be essentially none of the original dose left (0.2383^{23} is about 10^{-15}). So the maintenance level is the same as the original dose, 325 mg.

13. The simplified graph of the level of the drug in the bloodstream is concave down and increasing. Therefore the rate of increase in the early periods is larger than the rate of increase in the later periods. Therefore, if r_1 is the average rate of increase of the level over 10 days, r_2 the average rate of increase during the first five days, and r_3 is the average rate of increase during the last period of 5 days, we have $r_3 < r_1 < r_2$.

15. For $x_n = 4n$ we have: $x_1 = 4$, $x_2 = 8$, $x_3 = 12$, $x_4 = 16$, $x_5 = 20$, $x_6 = 24$.

17. For $x_n = (1/2)n$ we have: $x_1 = 0.5$, $x_2 = 1$, $x_3 = 1.5$, $x_4 = 2$, $x_5 = 2.5$, $x_6 = 3$.

19. For $x_n = n^3 - 10$ we have: $x_1 = -9$, $x_2 = -2$, $x_3 = 17$, $x_4 = 54$, $x_5 = 115$, $x_6 = 206$.

21. For $a_n = 2^n/3^n$ we have: $a_1 = 2/3$, $a_2 = 4/9$, $a_3 = 8/27$, $a_4 = 16/81$, $a_5 = 32/243$, $a_6 = 64/729$.

23.   For $x_n = 1/n$ we have: $x_1 = 1, x_2 = 1/2, x_3 = 1/3, x_4 = 1/4, x_5 = 1/5, x_6 = 1/6$.

25.   For $p_n = 1 - (0.2)^n$ we have $p_1 = 0.8, p_2 = 0.96, p_3 = 0.992, p_4 = 0.9984, p_5 = 0.99968, p_6 = 0.999936$.

27–31.   The sequences whose general term is provided in Problems 15–19 diverge to positive infinity since the size of the nth term grows indefinitely as n increases.

33.   The sequence $a_n = 2^n/3^n$ converges to 0. The term a_n may be rewritten as $(2/3)^n$, so it becomes evident that the fraction becomes smaller as n increases.

35.   The sequence $y_n = 1/n$ converges to 0 as n increases.

37.   The sequence $p_n = 1 - (0.2)^n$ converges to 1 as n increases.

39–41.   The points (n, x_n) are on a straight line, with slopes = 4, 3 and 1/2, respectively. In all three cases, the sequences are strictly increasing and exhibit no concavity.

43.   The points $(n, n^3 - 10)$ lie on the portion of the graph of the cubic polynomial $f(x) = x^3 - 10$ that is on the right of the vertical axis. The graph is both concave up and is strictly increasing for positive values of x.

45.   The points $(n, 2^n/3^n)$ lie on a graph that is strictly decreasing and concave up. The points of the sequence are approaching the line $y = 0$.

47.   The points $(n, 1/n)$ are on a graph that is strictly decreasing and concave up, corresponding to the branch of the hyperbola $y = 1/x, x > 0$.

49.   The points $(n, 1 - (0.2)^n)$ are on a graph that is strictly increasing and concave down.

51.   a.

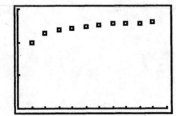

The scatter plot seems to be leveling off at a height of about 2.7.

b. $e_{100} = 2.7048138, e_{500} = 2.7155685, e_{1000} = 2.7169239, e_{10000} = 2.7181459,$

$e_{100000} = 2.7182682, e_{1000000} = 2.7182805$. The limit appears to be a number close to 2.71828. In fact the limit is the number e: $e = 2.718281828$

53.   The sequence $f_n = (1 - 1/n)^n$ has limiting value $1/e$.

55.   The limit of the sequence $(1 + n)^{1/n}$ is 1.

Section 5.2   Modeling with Difference Equations

1.   a.   According to the formula developed in part (b) of Example 3, the value of the account after 65 years is
$$B_{40} = 63,000(1.05)^{40} - 60,000 = \$383,519.$$

b.   If she started at 20, n is now 45 and the calculation is
$$B_{45} = 63,000(1.05)^{45} - 60,000 = \$506,056.$$

3. The population will die out when $0 = C_n = 40,000 - 10,000(1.20)^n$, that is, when $(1.20)^n = 4$. Since $(1.20)^7 = 3.58$ and $(1.20)^8 = 4.30$, the farmer will discover near the end of the 7th month that fewer than 8000 chickens are available to be harvested.

5. There are a variety of formats for these equations.
   a. A single deposit of $2000 earning 6% interest
      $b_{n+1} = (1.06)b_n$, $b_0 = 2000$
   b. An initial deposit of $2000 and yearly deposits of $1000 earning 6% interest
      $b_{n+1} = (1.06)b_n + 1000$, $b_0 = 2000$
   c. An initial deposit of $2000 and yearly deposits that increase by $1000, earning 6% interest.
      $b_{n+1} = (1.06)b_n + 2000 + 1000n$, $b_0 = 2000$
   d. An initial deposit of $2000 and yearly deposits that increase by 10%, earning 6% interest.
      $b_{n+1} = (1.06)b_n + 1000(1.1)^n$, $b_0 = 2000$

7. a. With W_n representing the population of bowhead whales, the difference equation model for the population under an annual harvest of 50 is $W_{n+1} = (1.03)W_n - 50$, with $W_{1992} = 8000$.
   b. $W_{1992} = 8000$, $W_{1993} = 8190$, $W_{1994} = 8386$, $W_{1995} = 8587$, $W_{1996} = 8795$, $W_{1997} = 9009$, $W_{1998} = 9229$, $W_{1999} = 9456$, and $W_{2000} = 9690$.
   c. To maintain a population of 8000 over the initial year, the Eskimos would harvest 3% of 8000. Thus the harvest level could be set at 240.

9. a. Letting r represent the proportion learned, the difference equation is $W_{n+1} = W_n + r(400 - W_n)$.
   b. For the above equation to model the process of learning, we require $0 < r < 1$.
   c. Answers will vary.
   d. For $r = 0.25$, we get the graph:

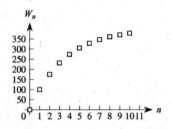

11. In this scenario, the numbers of pairs of rabbits for the first twelve months are 1, 1, 1, 2, 3, 4, 6, 9, 13, 19, 28, 41, 60. The difference equation that governs this growth is $P_{n+3} = P_{n+2} + P_n$.

13. For the difference equation $x_{n+1} = x_n + kn$, the first differences are $\Delta y = x_{n+1} - x_n = kn$, or $0, k, 2k, 3k, 4k, \dots$. The second differences are $\Delta^2 y = k, k, k, k, \dots$. Therefore, the solution is a quadratic polynomial.

Section 5.3   The Logistic or Inhibited Growth Model

In Exercises 1–4, data for a logistic model $P_{n+1} = (1+a)P_n - bP_n^2$ are given and the first ten values predicted by the model are requested, as is the limiting value for the population in each case.

1. With an initial population of 10, the first 10 estimates produced by the logistic model are 10.15, 10.30, 10.45, 10.61, 10.76, 10.92, 11.08, 11.24, 11.40, and 11.57. The limiting population is 40. The closest the estimates come to the limit during the first five time periods is within $40 - 10.76 = 29.24$, barely one-quarter of the way to the limit.

3. With an initial population of 5, the first 10 estimates produced by the logistic model are 5.10, 5.20, 5.30, 5.40, 5.51, 5.61, 5.72, 5.83, 5.95, and 6.06. The limiting population is 200. The closest the estimates come to the limit during the first five time periods is within $20 - 5.51 = 194.49$.

5. Since $a = 0.05$ and $b = 0.00002$, the limiting value in each case is $L = \dfrac{a}{b} = 2500$.

a.

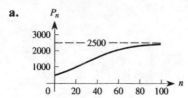

b.

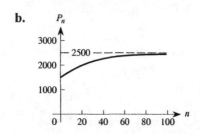

c.

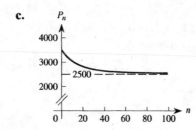

7. a.

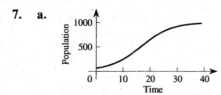

b. Given two points, an estimate can be produced for another point by using the line through the two points. The data tells us that between the 10th and 12th time periods, the population grew 50 units, an average of 25 units per time period. An estimate for the population at the 11th time period is $250 + 25 = 275$. Since the graph is concave up, the straight line segment from (10, 250) and (12, 300) lies above the curve, so this estimate is an overestimate.

c. The data tells us that between the 30th and the 31st periods, the population grows by 10 per day. An estimate for the 35th period is then $910 + 4(10) = 950$. Since the graph is concave up, the straight line through (30, 900) and (31, 910) lies above the curve at the 35th period, so this estimate is an overestimate.

9. With the estimates $a = 0.05$ and $b = 0.000003125$, we get the values 100, 104.97, 110.18, 115.65, and 121.39. These estimates are obtained by solving the system of equations $a/b = 16,000$, and $105 = (1 + a)100 - b\,(10,000)$.

11.   The table below summarizes the findings.

% change from 0.00002	10% inc	20% inc	30% inc	10% dec	20% dec
b	0.000022	0.000024	0.000026	0.000018	0.000016
L	2273	2234	1923	2778	3125
% change from 2500	9.1% inc	16.7% inc	23.1% inc	11% dec	25% dec

If a remains fixed, a change in b results in an inverse change in L. That is, if b increases, L decreases, and if b decreases, L increases. The percentage change in L is larger than that in b if b decreases, and is smaller if b increases.

13.   **a.**   The increased competition for the scarce resource of water will increase the value of b. If the drought is severe enough, it may also compromise the reproductive success of the species and lower the value of a.

  b.   If the population at the time of the draught was small compared to the original carrying capacity, the post-draught growth will follow the new model and approach a new smaller carrying capacity. If the population is larger than the original carrying capacity at the time of draught, there will be a rapid decrease in the population until it is near the new, smaller carrying capacity.

15.   **a.**   The following is the graph of $\Delta P_n = aP_n - bP_n^4$:

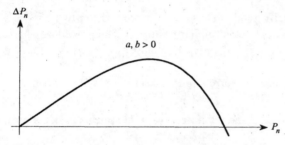

  b.   The limiting population L is the positive root of $0 = aP_n - bP_n^4 = P_n(a - bP_n^3)$. Thus $L = \sqrt[3]{\dfrac{a}{b}}$.

  c.   The population is increasing since $\Delta P_n > 0$ when $0 < P_n < L$.

  d.   The population is decreasing since $\Delta P_n < 0$ when $P_n > L$.

  e.   If P_0 is too large, the graph of ΔP_n will decrease only, so there is no point of inflection for P_n. If P_0 is small enough, the graph of ΔP_n will resemble that in part (a). Thus, ΔP_n clearly has a maximum value that corresponds to an inflection point for the solution of P_n.

17.

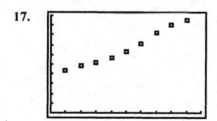

The formulas on the next page give the best-fitting function in each family: d stands for percentage of full bone density and t is age in years. The r-value for each fit is given at the right. The logistic model was found on a TI-83 calculator.

Logistic: $d = \dfrac{762.22}{1+19.16e^{-0.0573t}}$ $r = 0.9940$

Linear: $d = 3.425t + 32.528$ $r = 0.9864$

Exponential: $d = 38.37(1.0531)^t$ $r = 0.9939$

Power: $d = 29.30t^{0.3724}$ $r = 0.9462$

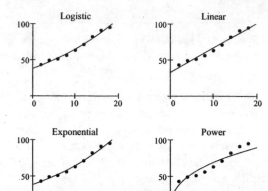

The logistic and exponential fits are about equally good. (These two functions differ by no more than 1 over the range of data, so we would expect the fits to have similar r-values.) The procedure of Example 7 would produce the logistic difference expression $\Delta d_n = 0.1387d_n - 0.00053d_n^2$. As in Problem 16, this produces a poor fit to the data, not surprising since the correlation between $\dfrac{\Delta d_n}{d_n}$ and d_n is only -0.21.

19. **a.** From the table, the maximum height is about 255 and the inflection point appears to occur around week 5.

b. The difference model is $\Delta h_n = 0.8771h_n - 0.00366h_n^2$, where h_n is the height after n weeks. The correlation between $\dfrac{\Delta h_n}{h_n}$ and h_n is -0.94.

c. The plot shows the heights of the sunflowers over 12 weeks (dots) and the heights predicted by our logistic model (boxes). The following table gives the numerical values for the data and the model.

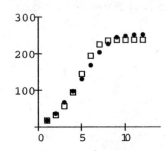

Data	17.9	36.4	67.8	98.1	131	169.5	205.3	228.3	247.1	250.5	238.8	254.4
Model	17.9	32.4	57.0	95.1	145.5	195.6	227.1	237.5	239.4	239.6	239.6	239.6

21. The difference model is $\Delta c_n = 1.2751c_n - 0.31488c_n^2$, where c_n is the concentration at time t_n. Note that the model assumes equal time intervals, which is nearly the case: the first interval is 22 units, which is 2 units shorter than the remaining intervals. For this data, the correlation between $\frac{\Delta c_n}{c_n}$ and c_n is -0.91.

23.

Data	9.6	29	71.1	174.6	350.7	513.3	594.4	640	655.9	661.8
Model	9.6	25.3	65.5	162.7	361.0	607.1	616.1	610.2	614.1	611.5

25. A decaying exponential fit for the transformed data is $\frac{\Delta P_n}{P_n} = 2.65(0.9931)^{P_n}$ which yields the nonlinear difference formula $\Delta P_n = 1.65P_n(0.9931)^{P_n}$. The table below shows the original data, the logistic model, and the model based on the exponential fit of the transformed data differences. The exponential model for the transformed data has correlation 0.97.

Data	9.6	29	71.1	174.6	350.7	513.3	594.4	640	655.9	661.8
Logistic	9.6	25.3	65.5	162.7	361.0	607.1	616.1	610.2	614.1	611.5
Exponential	9.6	33.4	103.8	238.5	361.2	441.1	497.4	540.5	575.4	604.6

27. **a.** The estimated logistic model is given by $\Delta E_n = 2.792E_n - 0.00923E_n^2$, where E_n is the electric capacity at period n, $n = 1$ corresponds to 1960, $n = 2$ corresponds to 1965, and so on. The correlation of the linear fit to the transformed differences is -0.85.

   b. According to this model, the limiting capacity is 303.5 gigawatts.

   c. From the table it appears that the capacity grew most rapidly between 1980 and 1985.

   d. The inflection point occurs when the capacity is half the maximum, that is, $\frac{1}{2} \cdot \frac{2.792}{0.00923} = 151.25$. Since the model gives the capacity in 1980 as 172.3, the inflection point will occur shortly before 1980.

29. $\Delta P_n = aP_n - bP_n^2$. The expression for ΔP_n is a quadratic which factors as $P_n(a - bP_n)$. The maximum occurs halfway between the roots, which are at $P_n = 0$ and $P_n = \frac{a}{b}$. Thus the maximum is at $P_n = \frac{1}{2} \cdot \frac{a}{b}$. At this point, $\Delta P_n = a\left(\frac{1}{2} \cdot \frac{a}{b}\right) - b\left(\frac{1}{2} \cdot \frac{a}{b}\right)^2 = \frac{1}{4} \cdot \frac{a^2}{b} = \frac{1}{4}b\left(\frac{a}{b}\right)^2 = \frac{1}{4}bL^2$.

Section 5.4 Newton's Laws of Cooling and Heating

1. If the initial temperature is $100°$ instead of $98.6°$, the difference equation modeling body temperature provides a solution of $T_n = 30(1+\alpha)^n + 70$. Note from Example 2, the computation of α does not involve the initial temperature, so $\alpha = 0.8356$. The time of death is determined from $30(0.8356)^n + 70 = 77.3$ providing $n = 7.87$ or 7 hours 52 minutes before 9:00. This is a 16-minute difference from the time of death found in Example 2.

3. Using the solution of the difference equation for heating with $R = 325$ and $T_0 = 40$, we have $T_n = 325 - 285(1 - \alpha)^n$. Solving $70 = 325 - 285(1 - \alpha)^{10}$ for $(1 - \alpha)$, we get the explicit form $T_n = 325 - 285(0.988939)^n$. The chicken will be ready in 60.8 minutes rather than 58.6 minutes.

5. If a can of soda at 70° is put into a freezer at 0° and it is determined that its temperature after ten minutes is 60°, then the time it takes for the temperature to drop to 40° is 36.3 minutes.

7. The solution to the difference equation governing Professor Smith's body temperature since the time of death is $T_n = (98.6 - 40)(1+\alpha)^n + 40 = 58.6(1+\alpha)^n + 40$. Solving $58.6(1+\alpha)^n + 40 = 67.3$ and $58.6(1+\alpha)^{n+1} + 40 = 63.1$ yields $1 + \alpha = (23.1/27.3) = 0.8462$. Substituting and solving for n gives $n = 4.574$, or 4 hours and 34 minutes before 9 A.M. Time of death was approximately 4:26 A.M.

9. The correlation coefficients for linear, exponential, and power fits are, respectively, –0.846504, –0.98412, and –0.990722, indicating the power fit $t = 1,814,899,979.11 \cdot T^{-5.196173}$ is the best fit. This model predicts that milk kept at 35° F will last 17.2 days.

11. The graph of the turkey's temperature is concave down and passes through the points (30, 70) and (60, 96). The straight line through these two points is $T = (13/15)t + 44$. Due to the concavity, a temperature at a time between 30 and 60 minutes will be possible only if it exceeds that predicted by the line. Outside of that range, a temperature will be possible only if it is smaller than that predicted by the line.
   a. $80 < (13/15)(45) + 44 = 83$, so $T(45) = 80$ is impossible.
   b. $85 > (13/15)(45) + 44 = 83$, so $T(45) = 85$ is possible.
   c. $105 < (13/15)(75) + 44 = 109$, so $T(75) = 105$ is possible.
   d. $115 > (13/15)(75) + 44 = 109$, so $T(75) = 115$ is impossible.

13. If t_1 is the time it takes for the temperature of a raw potato to go from 70° to 200°, and t_2 is the time it takes to go back to 70°, then we have $t_1 < t_2$.

15. a. The relevant information is shown in graphical form in the figure for part (b).
    b. Plotting ΔT versus $T - 8.6$ gives a nearly linear plot.

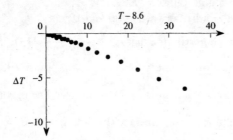

   c. $\Delta T_n = -0.1852(T_n - 8.6) + 0.0855 = -0.1852T_n + 1.6782 = -0.185(T_n - 9.0616)$
   d. We have $T_0 = 42.3$ (the first temperature entry in the table), $\alpha = 0.1852$, and $R = 9.0616$, so according to the solution on page 409, $T_n = 9.0616 + (42.3 - 9.0616)(1 - 0.1852)^n$ or $T_n = 9.0616 + 33.2384(0.8148)^n$. Here $n = 1$ correspond to the *second* entry in our table, that is, to the temperature at time 2.

Section 5.5   Geometric Sequences and Their Sums

For Problems 1 through 6 we use the formula $S_n = (1 - r^{n+1})/(1-r)$.

1. For $r = 0.5$, $S_{10} = 1.9990234375$, $S_{20} = 1.9999990463$, $S_{30} = 1.9999999991$; the partial sums are approaching 2.

3. For $r = 0.8$, $S_{10} = 4.57050327$, $S_{20} = 4.95388314$, $S_{30} = 4.99504824$; the partial sums are approaching 5.

5. For $r = 1.5$, $S_{10} = 170.99511719$, $S_{20} = 9973.77019024$, $S_{30} = 575251.17769865$. Notice that these values are not converging because $r > 1$.

7. If the number of new cases of the disease decreases by 20% each year, then the number of new cases in 1961 will be $6000(0.8)$, the number in 1962 will be $6000(0.8)^2$, and so on. The total number of cases between 1960 and 2000 will be

    $$6000 + 6000(0.8) + 6000(0.8)^2 + \cdots + 6000(0.8)^{40} = 6000\left(\frac{1 - 0.8^{41}}{1 - 0.8}\right),$$

    or 29,997 cases. The total number of cases from 1960 up to the year 2010 is $6000\left(\frac{1 - 0.8^{51}}{1 - 0.8}\right) = 30,000$. So, no more than 3 new cases will appear between 2000 and 2010.

9. If in 1980 the consumption of electricity in the U.S. was 2.5 billion kwh and usage has been growing 2% a year, then the amount of electricity used in the country between 1980 and 2000 is the sum of the first 21 terms of a geometric sequence with first term 2.5 billion and $a = 1.02$, a total of 64.4 billion kilowatt hours.

11. The sum of the first 19 terms of a geometric sequence with first term 70,600 and $a = 0.91$ is 653,722. This is our estimate of the total amount of rice in metric tons produced in the U.S. from 1984 through 2002.

13. We sum 15 terms of a series with first term 70,000 and $a = 1.08$. Approximately 1.901 million new pages of mathematical research were published from 1986 through 2000.

15. Based on a consumption growth rate of 2.5% and reserves of 2250 billion barrels, we will need to solve $26 \cdot \dfrac{1 - 1.025^{n+1}}{1 - 1.025} = 2250$, which is equivalent to $1.025^{n+1} = 3.16$. The solution is $n = 45.6$ years.

17. The distance traveled by the time the ball bounces an nth time, with an initial height of 12 ft and bouncing 3/4 of the way back up every bounce is given by the sum

    $$12 + 2 \cdot \left(\frac{3}{4}\right) \cdot 12 + 2 \cdot \left(\frac{3}{4}\right)^2 \cdot 12 + \cdots + 2 \cdot \left(\frac{3}{4}\right)^n \cdot 12$$

    $$= 12 + 2 \cdot \left(\frac{3}{4}\right) \cdot 12\left[1 + \frac{3}{4} + \cdots + \left(\frac{3}{4}\right)^{n-1}\right] = 12 + 18\left[\frac{1 - \left(\frac{3}{4}\right)^n}{1 - \frac{3}{4}}\right].$$

    If the ball were to bounce forever, the total distance covered would be

    $$12 + 18\left(\frac{1}{1 - \frac{3}{4}}\right) = 84 \text{ ft.}$$

19. The distance traveled by the time the ball bounces an nth time, with an initial height of 10 ft and bouncing 2/3 of the way back up every bounce is given by the sum

    $$10 + 2 \cdot \left(\frac{2}{3}\right) \cdot 10 + 2 \cdot \left(\frac{2}{3}\right)^2 \cdot 10 + \cdots + 2 \cdot \left(\frac{2}{3}\right)^n \cdot 10$$

    $$= 10 + 2 \cdot \left(\frac{2}{3}\right) \cdot 10\left[1 + \frac{2}{3} + \cdots + \left(\frac{2}{3}\right)^{n-1}\right] = 10 + \frac{40}{3}\left[\frac{1 - \left(\frac{2}{3}\right)^n}{1 - \frac{2}{3}}\right].$$

If the ball were to bounce forever, the total distance covered would be

$$10+\frac{40}{3}\left(\frac{1}{1-\frac{2}{3}}\right)=50 \text{ ft.}$$

The total distance covered has decreased by 20 feet when the bounce factor decreases from 3/4 to 2/3.

21. The sum is $\dfrac{25}{100}\left(1+\dfrac{1}{100}+\dfrac{1}{10000}+\cdots\right)=\dfrac{25}{100}\cdot\dfrac{1}{1-\dfrac{1}{100}}=\dfrac{25}{99}.$

23. The solution to $\dfrac{x_{n+1}}{x_n}=n^2$ will grow at a much faster rate than that of $\dfrac{x_{n+1}}{x_n}=n$ since each term is derived from the previous term by multiplying by n^2, a much larger number than n. The solution will be the square of the solution to Problem 22 (b), namely $[(n-1)!]^2 x_0$.

25. **a.** The best-fit exponential function is $G=105.23(0.7484)^n$, where G_n is weekly gross and n is the number of the week.

   b. The sum of the first 10 terms of a geometric sequence with general term $105.23(0.7484)^n$ is approximately 295.75, which gives an estimate of $296 million.

   c. The actual gross for the first 10 weeks was $302.9 million, so the estimate is quite good.

Chapter 5 Review Problems

1. **a.** $\{a_n\}=\{$ 5, 11, 17, 23, 29, ...$\}$ or $\{a_n\}=\{-1, 5, 11, 17, 23, ...\}$ for $n\geq 0$.
   b. $\{t_n\}=\{3, 9/2, 9, 81/4, 243/5, ...\}$
   c. $\{r_n\}=\{0.7, 0.91, 0.973, 0.9919, 0.99757, ...\}$ or $\{r_n\}=\{0, 0.7, 0.91, 0.973, 0.9919, ...\}$ for $n\geq 0$.

2. **a.** $\{2, 10, 18, 26, 34, ...\}$ **b.** $\{12, 4, -4, -12, -20, ...\}$

   c. $\{5, 5/3, 5/9, 5/27, 5/81, ...\}$ **d.** $\{10, 11, 8, 17, -10, ...\}$

3. **a.** $\{2, 7, 9, 16, 25, ...\}$ **b.** $\{3, 7, 10, 17, 27, ...\}$

4. The drug level follows the difference equation $D_{n+1}=0.4D_n+100, D_0=100$, where n is the number of six-hour periods. The solution is $D_n=(100/0.6)\left(1-0.4^{n+1}\right)$. Eight days represents 32 six-hour periods, and so the drug level will be $D_{32}=167$ mg.

5. The maintenance level of the drug is given by $B(1-0.30)=B/0.7$, where B is the periodic dosage, regardless of the initial dose. To achieve a maintenance level of 30 mg, the repeated dosage should be $(0.7)30=21$ mg.

6. There is a maintenance level given by $30/(1-0.80)=150$ lb.

7. The model is a logistic model $v_{n+1}=(1+a)v_n-bv_n^2$, where $a=0.3$ and $b=0.00002$. The limiting value, that is, the eventual number of VCR owners, is $a/b=15,000$. The point of inflection occurs when the number of VCRs is half this value, or 7500. By generating values of this sequence using $v_0=40$, we find the inflection point occurs in 22 years.

8. We have $a = 0.20$ and the limiting population $a/b = 12,000$, which gives an inhibiting constant of $b = 0.00001667$. The difference equation describing the population in year n is

$$P_{n+1} = 1.2P_n - 0.00001667(P_n)^2, \quad P_0 = 1000.$$

9. **a.** $P_{n+1} = 1.3P_n - 0.0004(P_n)^2$

   b. $P_{n+1} = 1.3P_n - 0.0004(P_n)^2 - 2000$

   c. $P_{n+1} = 1.3P_n - 0.0004(P_n)^2 + 400$

   d. $P_{n+1} = 1.3P_n - 0.0004(P_n)^2 - 0.10P_n$

10. **a.** The difference equation is $M_{n+1} = 0.85M_n$, $M_0 = 100\%$. A graph of the values over time is:

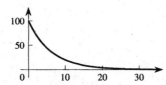

   The percentage in 2005 will be $100(0.85)^{35} = 0.34\%$.

   b. The difference equation is $P_{n+1} = 1.1P_n + 2000$, $P_0 = 50,000$. A graph of the values over time is:

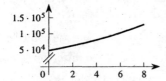

   The money invested in 2005 is found by computing the successive values of the sequence up to $P_9 = \$145,056.34$.

   c. If P_n and T_n are the budgets for print and television advertising, respectively, after n years from 1998, then the corresponding difference equations are

   $$P_{n+1} = P_n - 20,000, \quad P_0 = 2,000,000$$
   $$T_{n+1} = 1.1T_n, \quad T_0 = 2,000,000.$$

   A graph of the print media budget is:

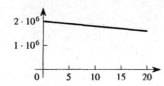

   A graph of the television budget is:

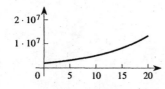

   At the end of 5 years, we have $P_5 = \$1,900,000$ and $T_5 = \$3,221,020$. The total advertising budget is growing at an almost exponential rate since the television budget is growing exponentially while the print media budget is decreasing linearly.

11.   **a.** Since the car loses 25% of the oil each week, the amount left at the end of week n is $0.75Q_n$, where Q_n is the amount at the beginning of the week. At the end of the week he adds 1 quart, so the difference equation is $Q_{n+1} = 0.75Q_n + 1$.

     b. Using the solution formula on page 368 with $a = 0.74$ and $B = 1$, we have $L = \dfrac{B}{1-a} = \dfrac{1}{1-0.75} = 4$ and the solution is $Q_n = 4 + 2(0.75)^n$.

     c. Solving $5 = 4 + 2(0.75)^n$ we find $n = 2.4$. Since this is not an integer, the level is never 5 quarts just after the weekly quart is added.

12. Increasing, concave up (Δ is positive and grows with n).

13. Increasing, concave up (Δ is positive and grows with n).

14. Increasing, concave down (Δ is positive and decreases with n).

15. Decreasing, concave down (Δ is negative and grows in absolute value with n).

16. Increasing, concave down (Δ is positive and decreases with n).

17. Increasing, concave up (Δ is positive and grows with n).

18. Δ changes sign, so the solution first increases, then decreases. When Δ is positive it is decreasing in size and when it is negative is increasing in size, so the solution is concave down everywhere.

19. Increasing, concave up (Δ is positive and grows with n). But note that if you compute the solution sequence on a calculator, once $3 \cdot (0.9)^n$ is less than the calculator's precision, all the differences will be computed as 12 and the solution becomes linear. (On a TI-83 Plus this happens around $n = 250$).

Chapter 6   Introduction to Trigonometry

Section 6.1   The Tangent of an Angle

1. a.

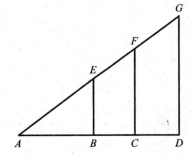

The measurements are in centimeters:

$AB = 2$ $AC = 3.1$ $AD = 4.3$
$AE = 2.5$ $AF = 3.8$ $AG = 5.3$
$BE = 1.4$ $CF = 2.2$ $DG = 3.1$

b. $\dfrac{AB}{AE} = 0.80$ $\dfrac{AC}{AF} = 0.82$ $\dfrac{AD}{AG} = 0.81$

$\dfrac{BE}{AB} = 0.70$ $\dfrac{CF}{AC} = 0.71$ $\dfrac{DG}{AD} = 0.72$

$\dfrac{BE}{AE} = 0.56$ $\dfrac{CF}{AF} = 0.58$ $\dfrac{DG}{AG} = 0.58$

c. The ratios that appear to be equal are grouped in rows in part (b). The ratios in each row are ratios of corresponding sides in similar triangles.

Problems 3–5 use the following figure:

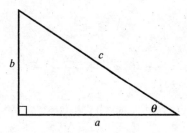

3. We can find b from the equation $\tan 16° = b/12$. This gives $b = 3.44094$. Now we can use the Pythagorean theorem, $c^2 = 12^2 + b^2$, to find $c = 12.4836$.

5. We can use the Pythagorean theorem, $30^2 = a^2 + 18^2$, to find $a - 24$. Then we can find θ from the equation $\tan \theta = 18/a$. This gives $\theta = 36.87°$.

7. Assuming that the flagpole rises vertically from perfectly flat horizontal ground, the height of the flagpole and the length of the flagpole's shadow are respectively, the side opposite and the side adjacent to the 28°- angle in a right triangle. Thus,

$$\frac{\text{height of the flagpole}}{50 \text{ ft}} = \tan 28° = 0.531709.$$

This gives 26.6 feet as the height of the flagpole.

9.

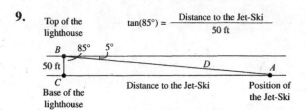

The right triangle *ABC* is formed by the position of the Jet-Ski, the top of the lighthouse, and the base of the lighthouse. Since the angle of depression from the top of the lighthouse to the Jet-Ski is 5°, the angle in triangle *ABC* is 85°. In triangle *ABC*, the indicated tangent equation gives $50 \tan 85° = 571.5$ ft as the distance from the base of the lighthouse to the Jet-Ski. We can find the straight line distance *D*, from the top of the lighthouse to the Jet-Ski by using the Pythagorean theorem, $D^2 = 50^2 + (571.5)^2$. This gives $D = 573.7$ ft.

11.

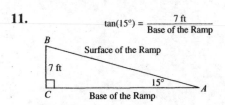

The right triangle *ABC* has the surface of the ramp as its hypotenuse, the base of the ramp as one side, with the height of the ramp forming the third side. The indicated tangent equation gives the length of the ramp's base as $\frac{7}{\tan 15°} \approx 26.1$ ft. Now using the Pythagorean theorem, we get $\sqrt{7^2 + (26.1)^2} \approx 27$ ft as the length of the ramp.

13. Using the equation

$$\tan 40° = \frac{\text{height of the cliff}}{24 \text{ ft}}$$

gives 20.1384 ft as the height of the cliff. this value differs by −0.8391 ft from the value obtained in Problem 12. The equation

$$\tan 40° = \frac{\text{height of the cliff}}{26 \text{ ft}}$$

gives the height of the cliff as 21.8166 ft, a value 0.83909 ft larger than that in Problem 12. The estimates of the height of the cliff range from 20.14 ft to 21.82 ft.

15. Seen from Jill's point of view, the situation can be described as follows:

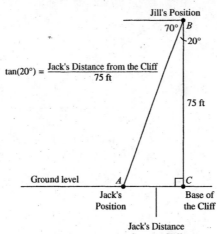

The indicated tangent equation puts Jack about $75\tan 20° \approx 27.3$ ft from the base of the cliff.

17. a.

$\tan\theta$	0	0.50	1.00	1.50	2.00	2.50	3.00
θ	0°	26.57°	45.00°	56.31°	63.43°	68.20°	71.57°

b.

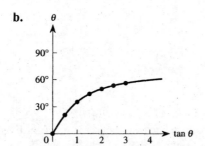

Section 6.2   The Sine and Cosine of an Angle

1.

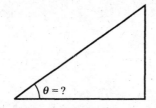

The bottom leg is 3.5 cm, the leg on the right is 2.5 cm, and the remaining side is 4.3 cm. We assume this is a right triangle. (We can check by using the converse of the Pythagorean theorem: $3.5^2 + 2.5^2 = 18.50$ and $4.3^2 = 18.49$, so our assumption is reasonable.)

$$\sin\theta = \frac{2.4}{4.3} = 0.5581 \qquad \cos\theta = \frac{3.5}{4.3} = 0.8140 \qquad \tan\theta = \frac{2.5}{3.5} = 0.7143$$

Using any one of these three ratios, we find that θ is about 33.54°.

Problems 3–7 use the following figure:

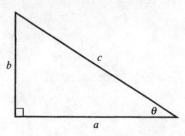

3. We can find *a* from the equation $\cos 16° = a/12$, and *b* from the equation $\sin 16° = b/12$. These yield $a = 11.5$ and $b = 3.3$, respectively. There are different computational approaches to this problem.

5. We can find θ from the equation $\sin \theta = 8/30$. This gives $\theta = 15.466°$. We can than compute *a* using the equation $\cos \theta = a/30$, which gives us $a = 28.9$.

7. We can find *c* from the equation $\cos 72° = 42/c$. This gives $c = 135.915$. We can now find the value of *b* from the Pythagorean theorem, $(135.916)^2 = 42^2 + b^2$. This gives $b = 129.263$.

9.

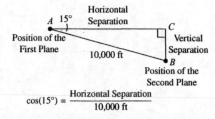

$$\cos(15°) = \frac{\text{Horizontal Separation}}{10,000 \text{ ft}}$$

The right triangle *ABC*, is formed by the line of sight between the two planes and their vertical and horizontal separations. The indicated sine equation gives a horizontal separation of $10,000 \cos 15° \approx 9659$ ft between the two planes.

11. **a.** The average angle of elevation θ satisfies the equation $\tan \theta = 2/20$. Thus, $\theta = 5.71059°$.

  b. The actual path upward of the airplane is first in a concave up pattern because the plane's height from the ground is increasing very rapidly, and then after a certain point, the plane's height levels off, indicating a concave down pattern.

  c. The straight line distance *c*, is given by $c^2 = 2^2 + 20^2$. Thus, $c = 20.0998$ mi.

13.

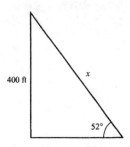

$$\frac{x}{400} = \sin 52° \implies x = \frac{400}{\sin 52°} = 507.6$$

They will cover 507.6 ft.

15.

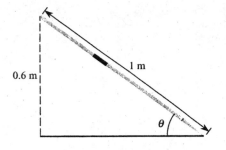

$$\frac{0.6}{1} = 0.6 = \sin\theta \implies \theta = \arcsin(0.6) \approx 36.9°$$

17.

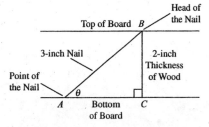

If the end of the nail just touches the bottom of the piece of wood, it will form the hypotenuse of the triangle *ABC*, whose altitude is the thickness of the wood and whose base is part of the bottom of the piece of wood. The angle formed by the nail and the bottom surface of the wood is the same as the angle formed by the nail and the top surface of the wood. Thus, θ must satisfy $\sin\theta = 2/3$. We can now calculate that $\theta = 41.81°$.

19.

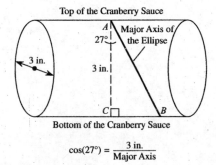

$$\cos(27°) = \frac{3 \text{ in.}}{\text{Major Axis}}$$

In the right triangle *ABC*, the hypotenuse is the major axis of a slice from the center of the roll of cranberry sauce. Using the indicated cosine function, we can calculate the length of that major axis as $3/\cos 27° \approx 3.37$ inches.

21. $\dfrac{\text{height}}{90} = \sin 72° \implies \text{height} = 90 \sin 72° = 85.6.$

The ladder can reach to a height of 85.6 feet.

23.

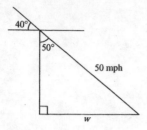

Now we have $w = 50\sin 50° = 38.3$. The speed to the east is 38.3 mph.

25. **a.**

$\cos\theta$	0	0.2	0.4	0.6	0.8	1
θ	90°	78.46°	66.42°	53.13°	36.87°	0°

b.

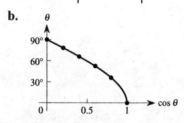

c. The inverse cosine function, $\arccos(x)$.

Section 6.3   The Sine, Cosine, and Tangent in General

1. $\sin 225° = -\sin 45° = -0.707$

3. $\tan 135° = -\tan 45° = -1$

5. $\sin 330° = -\sin 30° = -0.5$

7. $\cos 315° = \cos 45° = 0.707$

9. $\cos 840° = \cos(840° - 720°) = \cos 120° = -\cos 60° = -0.5$

11. $\cos(-240°) = \cos(-240° + 360°) = \cos 120° = -\cos 60° = -0.5$

13. $300°$ is in Quadrant IV, so $\sin 300°$ is negative.

15. $240°$ is in Quadrant III, so $\cos 240°$ is negative.

17. $300°$ is in Quadrant IV, so $\tan 300°$ is negative.

19. $270°$ is on the negative y-axis, so $\cos 270°$ is zero.

21. $215°$ is in Quadrant III, so $\sin 215°$ is negative.

23. $520°$ is in Quadrant II, so $\cos 520°$ is negative.

25. $925°$ is in Quadrant III, so $\tan 925°$ is positive.

27. $1000°$ is in Quadrant IV, so $\cos 1000°$ is positive.

29. $-480°$ is in Quadrant III, so $\sin(-480°)$ is negative.

31. $-500°$ is in Quadrant III, so $\tan(-500°)$ is positive.

33. a. $f(x) = \sin x \cos x$ is positive between 0° and 90°, negative between 90° and 180°, and negative between 450° and 540°.

b. $f(x) = 0$ where either sine or cosine is 0, that is, at $x = 0°$, 90°, 180°, 270°, 360°, 450°, and 540°.

c.

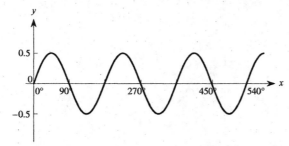

d. The function appears to be periodic with period 180°.

35. a.

$\cos\theta$	1	0.75	0.5	0.25	0	−0.25	−0.5	−0.75	−1
θ	0°	41.41°	60°	75.522°	90°	104.478°	120°	138.59°	180°

b.

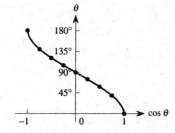

c. This curve is the graph of the inverse cosine function, arccos(x).

Section 6.4   Relationships among Trigonometric Functions

1. Since we know that $\sin\theta = 0.3$, we can use the Pythagorean theorem, $(\cos\theta)^2 + (0.3)^2 = 1$, to determine that $\cos\theta = \sqrt{1 - 0.09} = 0.9539$. Then $\tan\theta = \sin\theta / \cos\theta = 0.3/0.9539 = 0.3145$. Using the sine information, $\theta = \arcsin(0.3) = 17.46°$

3. $\sin\theta = \dfrac{3}{5}$, $\cos\theta = \sqrt{1 - \left(\dfrac{3}{5}\right)^2} = \sqrt{\dfrac{16}{25}} = \dfrac{4}{5}$, $\tan\theta = \dfrac{\sin\theta}{\cos\theta} = \dfrac{3/5}{4/5} = \dfrac{3}{4}$

5. Sketch a right triangle with legs 3 and 4 units long; the hypotenuse will have length 5. Since the shorter leg is opposite angle θ, $\sin\theta = \frac{3}{5}$ and $\cos\theta = \frac{4}{5}$.

7. $(0.52)^2 + (\cos\theta)^2 = 1$, so $\cos\theta = \pm\sqrt{1 - (0.52)^2} = \pm 0.8542$. Since the angle is in Quadrant II, the cosine is negative, so $\cos\theta = -0.8542$, and

$$\tan\theta = \frac{\sin\theta}{\cos\theta} = \frac{0.52}{-0.8542} = -0.6088.$$

9. The cosine is positive in Quadrant IV, so $\cos\theta = \sqrt{1 - (0.7)^2} = 0.7141$, and

$$\tan\theta = \frac{\sin\theta}{\cos\theta} = \frac{-0.7}{0.7141} = -0.9803.$$

11.   **a.** The graphs for (i) and (ii) are shown below; Graph (ii) is an identity.

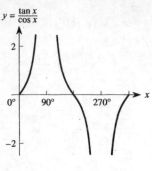

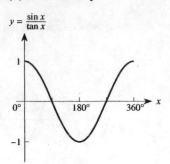

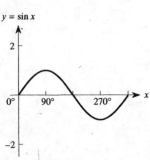

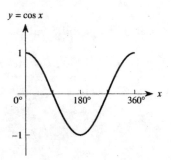

b. $\dfrac{\sin x}{\tan x} = \dfrac{\sin x}{\dfrac{\sin x}{\cos x}} = \sin x \cdot \dfrac{\cos x}{\sin x} = \cos x$

c. $x = 0°$ and $x = 180°$ both satisfy Equation (i).

Exercising Your Algebra Skills

1. $\cos x \tan x = \cos x \cdot \dfrac{\sin x}{\cos x} = \sin x$

2. $(1 - \sin x)(1 + \sin x) = 1 - \sin^2 x = \cos^2 x$

3. $(1 - \cos x)(1 + \cos x) = 1 - \cos^2 x = \sin^2 x$

4. $(\sin\theta + \cos\theta)^2 = \sin^2\theta + 2\sin\theta\cos\theta + \cos^2\theta = (\sin^2\theta + \cos^2\theta) + 2\sin\theta\cos\theta = 1 + 2\sin\theta\cos\theta$

5. $(\sin\theta - \cos\theta)^2 = \sin^2\theta - 2\sin\theta\cos\theta + \cos^2\theta = (\sin^2\theta + \cos^2\theta) - 2\sin\theta\cos\theta = 1 - 2\sin\theta\cos\theta$

6. $\cos^3 x + \sin^2 x \cos x = \cos x\left(\cos^2 x + \sin^2 x\right) = \cos x(1) = \cos x$

7. $\cos\theta + \tan^2\theta\cos\theta = \cos\theta\left(1 + \tan^2\theta\right) = \cos\theta \cdot \dfrac{1}{\cos^2\theta} = \cos\theta$

8. $\tan^2\theta - \dfrac{1}{\cos^2\theta} = \tan^2\theta - \left(1 + \tan^2\theta\right) = \tan^2\theta - 1 - \tan^2\theta = -1$

9. $\left(1 - \dfrac{1}{\cos x}\right)\left(1 + \dfrac{1}{\cos x}\right) = 1 - \dfrac{1}{\cos^2 x} = 1 - \left(1 + \tan^2 x\right) = 1 - 1 - \tan^2 x = -\tan^2 x$

Section 6.5   The Law of Sines and the Law of Cosines

1. We immediately have $C = 180° - 26° - 63° = 91°$. Using the law of sines, we have

$$\frac{\sin 63°}{12} = \frac{\sin 26°}{a} \quad \text{and} \quad \frac{\sin 63°}{12} = \frac{\sin 91°}{c}.$$

Solving these equation, we get $a = 5.90$ and $c = 13.47$.

3. We immediately have $C = 180° - 35° - 65° = 80°$. Using the law of sines, we have

$$\frac{\sin 80°}{24} = \frac{\sin 35°}{a} \quad \text{and} \quad \frac{\sin 80°}{24} = \frac{\sin 65°}{b}.$$

Solving these equation, we get $a = 13.98$ and $b = 22.09$.

5. Since $A = 40°$ and $a = 10$, we can use the law of sines and $b = 12$ to find B. This gives us

$$\frac{\sin 40°}{10} = \frac{\sin B}{12} \implies \sin B = 0.7713451316.$$

There are two possible values of B with that sine value: $B = 50.47°$ and $B = 180° - 50.47° = 129.53°$. The first value for B gives $C = 180° - 40° - 50.47° = 89.53°$. Using the law of sines again, we get $\sin 40°/10 = \sin 89.53°/c \implies c = 15.56$. The other value for B gives $C = 180° - 40° - 129.53° = 10.47°$, which gives $\sin 40°/10 = \sin 10.47°/c \implies c = 2.83$.

7. There are a great many good ways to answer the question "Where is the third ship?" One way would be to give the heading from the southernmost ship, which is $41°$ north of east and the distance indicated as b in the following diagram.

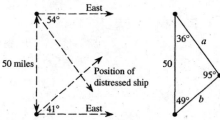

From the law of sines we get

$$\frac{\sin 95°}{50} = \frac{\sin 36°}{b} \implies b = \frac{50 \sin 36°}{\sin 95°} = 29.5 \text{ miles.}$$

9. **a.** Problem 4 represents the case $b < a$, where $b = 6$ and $a = 10$. The following diagram shows that only one triangle can be formed: $AC = 6$ is less than the radius $BC = 10$, which intersects the ray AB exactly once.

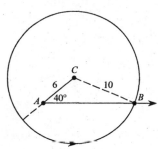

b. Problem 5 represents the case $b > a$, where $b = 12$ and $a = 10$ (and 12 is somewhat larger than 10). The following diagram shows that two different triangles can be formed: $AC = 12$ is greater than the radius $BC = 10$, but BC is short enough to intersect the ray AB at two different points.

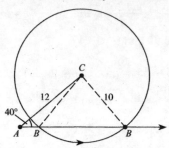

c. Problem 6 represents the case $b > a, s$ where $b = 18$ and $a = 10$ (and 18 is much larger than 10). The following diagram shows that no triangle can be formed: $AC = 18$ is greater than the radius, but it exceeds the radius too much so that side BC cannot intersect the ray AB.

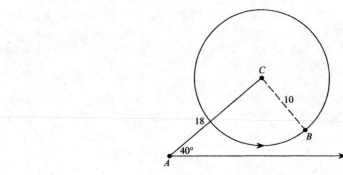

11.

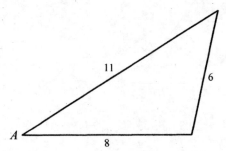

The angle that we are trying to find is A. By the law of cosines,
$$6^2 = 8^2 + 11^2 - 2 \cdot 8 \cdot 11 \cdot \cos A \implies \cos A = 149/176 = 0.8466 \implies A = 32.2°.$$

13.

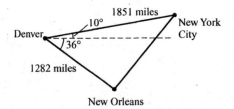

Call the New York–New Orleans distance d, we have by the law of cosines,
$$d = \sqrt{1851^2 + 1282^2 - 2 \cdot 1851 \cdot 1282 \cos(36° + 10°)} = 1332 \text{ miles.}$$

15.

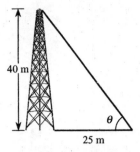

We could find the length L of the guy wires using the Pythagorean theorem, but instead we will first find the angle θ and then use θ to find the length.

$$\tan\theta = \frac{40}{25} = 1.6 \Rightarrow \theta = 58°$$

Then $\dfrac{40}{L} = \sin\theta = \sin 58° = 0.848 \Rightarrow L = \dfrac{40}{0.848} = 47.2$ m.

17.

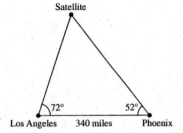

We will first solve for the vertical distance h between the satellite and Earth. Using the two right triangles formed by dropping a perpendicular from the satellite to the Earth, we have

$$\frac{h}{\tan 52°} + \frac{h}{\tan 72°} = 340 \Rightarrow h = 340 \cdot \frac{1}{\dfrac{1}{\tan 52°} + \dfrac{1}{\tan 72°}} = 307.4 \text{ miles.}$$

The distance d between Phoenix and the satellite is the hypotenuse of a right triangle, so we can find d:

$$\frac{307.4}{d} = \sin 52° = 0.7880 \Rightarrow d = \frac{307.4}{0.7880} = 390 \text{ miles}$$

19.

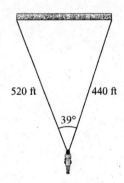

From the law of cosines we have:

$$\text{length} = \sqrt{520^2 + 440^2 - 2 \cdot 520 \cdot 440 \cdot \cos 39°} = 329 \text{ meters.}$$

21.

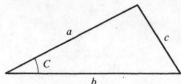

Using the law of cosines, we get $c = \sqrt{3^2 + 5^2 - 2 \cdot 3 \cdot 5 \cdot \cos 20^\circ} = 2.41$.

Chapter 6 Review Problems

1. a.

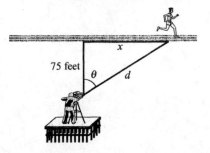

$$\frac{75}{d} = \cos\theta \Rightarrow d = \frac{75}{\cos\theta}$$

b. If $d = 240$, then $\cos\theta = 75/240 = 0.3125 \Rightarrow \theta = 71.79^\circ$. If the camera picks up the runner opposite the platform, then the interval of angles would be 0° to 72°. If the camera picks up the runner as soon as possible, the interval of angles would be -72° to 72°.

c. $-72^\circ \leq \theta \leq 72^\circ$

2. a. The distance the runner covers is x in the diagram for Problem 1.

$$\frac{x}{75} = \tan\theta \Rightarrow x = 75\tan\theta$$

b. For $\theta = 25^\circ$, $x = 75\tan 25^\circ = 35$ feet.

c. To find the angle corresponding to 150 feet, we solve: $150 = 75\tan\theta \Rightarrow \tan\theta = 2 \Rightarrow \theta = \arctan(2) = 63.4^\circ$. So, the camera pans from 0° to 63.4°.

3.

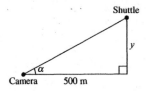

a. $\dfrac{y}{500} = \tan\alpha \Rightarrow y = 500\tan\alpha$

b. The height when $\alpha = 20^\circ$ is $500\tan 20^\circ = 182$ meters.

c. The height when $\alpha = 40^\circ$ is $500\tan 40^\circ = 419.5$ meters.

d. Solving $2000 = 500\tan\alpha$ for α, we get $\tan\alpha = 4 \Rightarrow \alpha = \arctan(4) = 76^\circ$.

4. a.

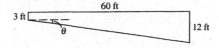

The angle of depression is labeled θ in the diagram.

$$\tan\theta = \frac{12-3}{60} = 0.150 \Rightarrow \theta = \arctan(0.150) = 8.5°$$

b. Taking the water surface at one corner of the shallow end as the origin, the equation would be
$y = -0.15x - 3$.

c. The length of the pool is 60 feet. The width of the pool is given as 25 feet, so if we let the bottom coincide with the xy-plane, the slope of a diagonal line across the bottom would be $-25/60 = -5/12$, and the equation would be $y = -0.417x$.

5.

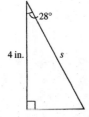

The longest length s is related to the angle of the slice by $\dfrac{4}{s} = \cos 28° \Rightarrow s = \dfrac{4}{\cos 28°} = 4.5$ inches.

6.

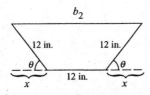

a. If 12 inches are bent up at $35°$, the depth d is given by $\dfrac{d}{12} = \sin 35° \Rightarrow d = 12\sin 35° = 6.9$ inches.

b. For an angle of $55°$ the depth is $d = 12\sin 55° = 9.8$ inches.

7. From Problem 6 we know the heights (depths) corresponding to angles of $35°$ and $55°$, and b_1 is always 12, so we need to find the length labeled b_2 in the diagram for Problem 6 above.

a. For bend angles of $35°$, $b_2 = 12 + 2x$ and $\dfrac{x}{12} = \cos 35° \Rightarrow x = 12\cos 35° = 9.8$. Thus,

$b_2 = 12 + 2(9.8) = 31.6$, and the volume is $\frac{1}{2}(12 + 31.6)(6.9)(96) = 14,440.3$ cubic inches.

b. For bend angles of $55°$, $b_2 = 12 + 2x$ and $\dfrac{x}{12} = \cos 55° \Rightarrow x = 12\cos 55° = 6.9$. Thus,

$b_2 = 12 + 2(6.9) = 25.8$, and the volume is $\frac{1}{2}(12 + 25.8)(9.8)(96) = 17,781.1$ cubic inches.

c. From Problem 6 and the solving process for parts (a) and (b), we can derive the following formula:

$$\tfrac{1}{2}[12(12+24\cos\theta)](12\sin\theta)(96)=13,824(1+\cos\theta)\sin\theta$$

d. Graphing $(1+\cos\theta)\sin\theta$, we find a maximum of about 1.3 at $\theta=60°$, so the maximum volume is approximately $13,824(1.3)=17,971$ cubic inches.

8. It is easiest to solve part(b) first: We find the height h of the window using $\tan23°=h/220$. Thus $h=220\tan23°=93.4$ feet. Then H is h plus the remaining height of the building opposite, which is $220\tan36°=159.8$ feet.
   a. $H=93.4+159.8=253.2$ feet
   b. $h=93.4$ feet

9.

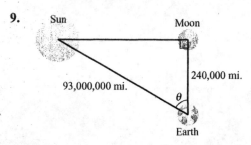

$$\cos\theta=\frac{240,000}{93,000,000}=0.00258 \Rightarrow \theta=\arccos(0.00258)=89.85°$$

10.

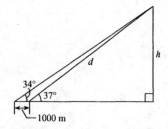

We will first use the law of sines to find the distance d from the second observation point to the top of the mountain. In the very narrow triangle the angle at the mountain top is $3°$ (the difference between the two observed angles of elevation).

$$\frac{d}{\sin34°}=\frac{1000}{\sin3°} \Rightarrow d=\frac{1000\sin34°}{\sin3°}=10,685 \text{ meters}$$

Then using the inner right triangle we have:

$$\frac{h}{10,685}=\sin37° \Rightarrow h=10,685\sin27°=6430 \text{ meters}$$

11. a. In 5 minutes the minute hand moves through an angle of $360/12=30°$. The vertical rise is thus $10\sin30°=5$ inches.
   b. At 50 minutes past the hour, the hand is already halfway (vertically) to the top, so in any time less than 10 minutes, it will move less than 5 inches vertically.
   c. 10 to 15 minutes after the hour, 15 to 20 minutes after the hour (going down), and 40 to 45 minutes after the hour (going up).

12. a. Using the identity $1+\tan^2\theta=\dfrac{1}{\cos^2\theta}$, we have:

$$\cos\theta=\sqrt{\frac{1}{1+\tan^2\theta}}=\sqrt{\frac{1}{1+1.20^2}}=0.640$$

Then, using the identity $\cos^2\theta + \sin^2\theta = 1$, we get:

$$\sin\theta = \sqrt{1-\cos^2\theta} = \sqrt{1-0.640^2} = 0.768$$

b.

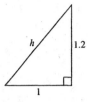

$h = \sqrt{1^2 + 1.2^2} = \sqrt{2.44} = 1.562$. Thus, we have:

$$\cos\theta = \frac{1}{h} = \frac{1}{1.562} = 0.640 \quad \text{and} \quad \sin\theta = \frac{1.2}{1.562} = 0.768$$

13.

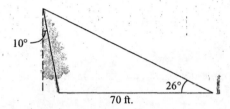

70 ft.

a. The angle at the base of the tree is $100°$, so the angle at the top is $180° - 100° - 26° = 54°$. Then by the law of sines, if h is the height of the tree (that is, the length of the trunk from bottom to top), we get:

$$\frac{70}{\sin 54°} = \frac{h}{\sin 26°} \Rightarrow h = \frac{70\sin 26°}{\sin 54°} = 37.9 \text{ feet}$$

The height of the top above the ground is $37.9\cos 10° = 37.3$ feet.

b. If the tree is leaning *toward* the observer, the calculations look like this:

$$h = \frac{70\sin 26°}{\sin 74°} = 31.9 \text{ feet}$$

So, the height of the top above the ground is $31.9\cos 10° = 31.4$ feet.

14.

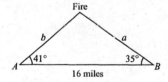

The angle between the two observers as seen from the fire is $180° - 41° - 35° = 104°$. Using the law of sines, we get:

$$\frac{b}{\sin 35°} = \frac{a}{\sin 41°} = \frac{16}{\sin 104°}$$

Therefore,

$$b = \frac{16\sin 35°}{\sin 104°} = 9.5 \text{ miles} \quad \text{and} \quad a = \frac{16\sin 41°}{\sin 104°} = 10.8 \text{ miles.}$$

15.

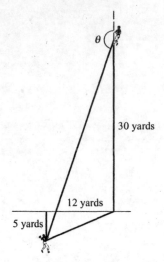

a. This distance is the hypotenuse of a triangle with legs 5 yards and 12 yards, which is 13 yards.

b. This angle has tangent $5/12$, so the angle is $\arctan(5/12) = 22.6°$.

c. The distance the ball travels to the receiver is $\sqrt{12^2 + (30+5)^2} = 37$ yards.

d. The angle θ through which the receiver must turn is the supplement of $\arctan(12/35)$, that is, $180° - 18.9° = 161.1°$.

16.

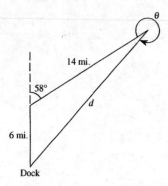

a. Using the law of cosines, we can express the distance d as

$$d = \sqrt{6^2 + 14^2 - 2 \cdot 6 \cdot 14 \cdot \cos 122°} = 17.9 \text{ miles.}$$

b. The acute angle α at the point where the motorboat turns is given by

$$\frac{\sin \alpha}{6} = \frac{\sin 122°}{17.9} \Rightarrow \sin \alpha = \frac{6 \sin 122°}{17.9} = 0.284 \Rightarrow \alpha = \arcsin(0.284) = 16.5°.$$

Then the angle θ through which the motorboat turns is $360° - \alpha = 163.5°$.

c. The time for the return trip is 17.9 miles/18 mph $\approx$ 1 hour. For the whole trip, add $(6+14)$ miles/18 mph $= 1.1$ hour, for a total of 2.1 hours.

d. The total cost is gallons used times price, or

$$\frac{(6+14+17.9) \text{ miles}}{6 \text{ mpg}} \cdot \frac{\$2.85}{\text{gallon}} = \$18.$$

17.

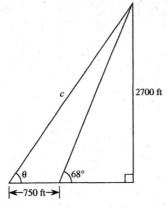

First we find the distance x from the base of the mountain to a point directly under the summit:

$$x = \frac{2700}{\tan 68°} = 1090.1 \text{ miles.}$$

Then

$$\tan\theta = \frac{2700}{750+x} = \frac{2700}{1840.1} = 1.467 \Rightarrow \theta = \arctan(1.467) = 55.7°.$$

The length of the cable c, could be found via the Pythagorean theorem, or as follows:

$$\frac{2700}{c} = \sin 55.7° \Rightarrow c = \frac{2700}{\sin 55.7°} = 3268 \text{ feet.}$$

18.

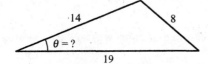

a. Using the law of cosines, we get:

$$8^2 = 19^2 + 14^2 - 2 \cdot 19 \cdot 14 \cdot \cos\theta \Rightarrow \cos\theta = \frac{493}{532} = 0.927 \Rightarrow \theta = \arccos(0.927) = 22°.$$

b. The height of the triangle is $14\sin 22° = 5.2$ cm.

c. The area is $\frac{1}{2}(19)(5.2) = 49.4$ cm^2.

d. If the angle θ is opposite side c, then a formula for the area is $\frac{1}{2}ab\sin\theta$.

Chapter 7 Modeling Periodic Behavior

Section 7.1   Introduction to the Sine and Cosine Functions

1.   The length of Janis's fingernails is a periodic function with a period of 1 week.

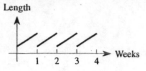

3.   The figure illustrates the graphing process for angles between $0°$ and $90°$.

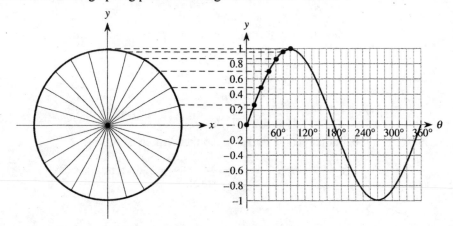

In Problem 5, to compare the measure of an angle θ in degrees to its measure in radians we use the proportion $\dfrac{t}{\pi} = \dfrac{\theta°}{180°}$.

5.   **a.** $3\pi/4$ radians corresponds to $135°$ **b.** $4\pi/5$ radians corresponds to $144°$
   c. $2\pi/3$ radians corresponds to $120°$ **d.** 1.5 radians corresponds to $86°$
   e. 2.5 radians corresponds to $143°$ **f.** 3 radians corresponds to $172°$
   g. $\pi/8$ radians corresponds to $22.5°$ **h.** $5\pi/3$ radians corresponds to $300°$
   i. $-3\pi/2$ radians corresponds to $-270°$ **j.** $-5\pi/3$ radians corresponds to $-300°$

7.   **a.** $f(30°) = \dfrac{\sqrt{3}}{2}$ **b.** $f(45°) = 1$ **c.** $f(120°) = -\dfrac{\sqrt{3}}{2}$

   d. $f(225°) = 1$ **e.** $f\left(\dfrac{\pi}{3}\right) = \dfrac{\sqrt{3}}{2}$ **f.** $f\left(\dfrac{\pi}{12}\right) = \dfrac{1}{2}$

   g. $f\left(\dfrac{3\pi}{8}\right) = \dfrac{\sqrt{2}}{2}$ **h.** $f\left(\dfrac{2\pi}{7}\right) = 0.9749$

9.

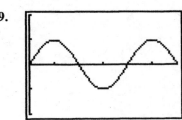

The two graphs are the same, since for any x,

$\sin x = \cos\left(x - \dfrac{\pi}{2}\right)$: the sine graph is the same as the cosine

graph shifted $\pi/2$ units to the right.

11. **a.** The period for the population of lynxes is roughly 10 years.
    b. The period for the population of hares is also roughly 10 years.
    c. The lynx population reached its maximum values in the years: 1847, 1857, 1867, 1877, 1886, 1895, 1906, 1915, 1927, and 1936. The lynx population was at a minimum in the years: 1852, 1862, 1872, 1881, 1891, 1901, 1908, 1920, and 1930.
    d. The hare population reaches its maximum values in 1853, 1857, 1861, 1864, 1873, 1876, 1886, 1896, 1904, 1913, 1923, and 1933. The hare population is at a minimum value in 1847, 1855, 1859, 1882, 1899, 1908, 1918 and, 1928.
    e. The years in which the population achieves a maximum (minimum) is separated by ten years from the next maximum (minimum).
    f. The years in which the hare population passed an inflection point are roughly half way between a maximum year and the following minimum year or half way between a minimum year and the following maximum year.

Section 7.2   Modeling Periodic Behavior with the Sine and Cosine Functions

1. **a.** The graph depicts a periodic function with period 2.
   b. The graph depicts a periodic function of period 2.
   c. The graph depicts a periodic function of period 2.
   d. The graph depicts a periodic function of period 1.
   e. The graph depicts a periodic function of period 2.
   f. The graph is not periodic.
   g. The graph is not periodic.
   h. The graph depicts a constant function. Constant functions can be considered periodic, but a definite period cannot be determined.
   i. The graph depicts a periodic function of period 2.
   j. The graph depicts a periodic function of period 20.

3. **a.** The amplitude of H is 3.6 hours.
   b. The period of H is 365 days.
   c. The length of the shortest day of the year is $12 - 3.6 = 8.4$ hours.
   d. The length of the longest day is $12 + 3.6 = 15.6$ hours.

5. From the text we know that the number of hours of daylight (from sunrise to sunset) in San Diego is given by $S(t) = 12 + 2.4 \sin\left(\frac{2\pi}{365}(t - 80)\right)$. Provided we are prepared to interpret "dark" as including twilight (from sunset to sunrise), the required formula is $H(t) = 24 - S(t) = 12 - 2.4 \sin\left(\frac{2\pi}{365}(t - 80)\right)$.

7. Tides are reasonably modeled by formulas of the form $D + A \sin(B(x - C))$. For the Bay of Fundy we are given information to conclude that the average tide level is $D = 25$, the amplitude is $A = 25$, and the frequency is $B = 2\pi/11$. We are not given times for the low tide and high tide, so we cannot determine a phase shift value. Thus any model of the form $25 + 25 \sin((2\pi/11)(x - C))$ would be appropriate. For simplicity, we could say $C = 0$ and get $W(t) = 25 + 25 \sin\left(\frac{2\pi}{11}t\right)$.

9. **a.** A sinusoidal model of an ocean wave will be of the form $A \sin(Bt)$. If the height of the wave is 4 feet, the amplitude of the model must be 2. Using our rule of thumb, the length of a four-foot wave would be 80 feet, making the period of our model 80. Using these observations, the model becomes $2 \sin\left(\frac{\pi}{40}t\right)$.

b. Reasoning as in part (a), the sinusoidal model will have an amplitude of 7.5 and a period of 300, so the model is $7.5 \sin\left(\frac{\pi}{150}t\right)$.

11. **a.** Once the oven has reached the target temperature, a function such as $T = 350 + 10\sin\left(2\pi\frac{t}{10}\right)$ might be an appropriate model, where T is temperature and t is time in minutes, and assume that "cycle" means full cycle from on to off to on again.

b. If we put the turkey in the oven at 350°, its average temperature will rise to equal the average temperature of the oven and then cycle (at some lag from the oven temperature). If the oven temperature were constant at 350°, Newton's law of heating would give the turkey temperature as $T_{turkey} = 350 - 310e^{-t/175}$. Since the oven is turning on and off, the temperature will rise faster and some times than others, and once the average temperature has reached 350°, the turkey temperature will cycle through a much narrower range than the oven temperature. If it takes 60 minutes for the turkey to reach 130°, the turkey's time constant is so large (that is, the turkey responds so slowly to changes in the surrounding temperature) that the temperature fluctuations in the turkey will be less than 0.1°. In fact we'll take the turkey out of the oven long before it reaches the minimum temperature of 340°, which would happen in about 10 hours.

13. **a.** The period is $2(20-8) = 24$ and the amplitude is $\frac{72-30}{2} - 21$. The vertical shift is the average of the min and max, or $\frac{30+72}{2} = 51$. A sine function with these parameters is $51 + 21\sin\left(\frac{2\pi}{24}(x-C)\right)$. To find a possible value for C, note that if C were 0 the first maximum would occur one quarter of the way through the period, at $x = 6$. We want this peak to be at 8 instead, so $C = 2$ and the final formula is

$$f(x) = 51 + 21\sin\left(\frac{2\pi}{24}(x-2)\right).$$

b. The cosine parameters will be the same except for C. With $C = 0$ the first peak is at $x = 0$, so we need to shift this peak 8 units to the right. Thus $C = 8$ and the final formula is

$$f(x) = 51 + 21\cos\left(\frac{2\pi}{24}(x-8)\right).$$

15. Assuming that the mean temperature is the average of the high and the low we have a mean of $\frac{64+(-20)}{2} = 22$ and an amplitude of $\frac{64-(-20)}{2} = 42$. The period is roughly 365 days, and if we use a sine function we must shift the minimum to the right by one quarter of a period plus 40 more days, that is, by $\frac{365}{4} + 40 = 131.25$ days. A good approximation would be $T(t) = 22 + 42\sin\left(\frac{2\pi}{365}(t-131)\right)$.

17. Since trigonometric functions are periodic there are infinitely many correct formulas for each of the graphs shown in the text. For each graph there are two infinite families of formulas, one family based on the sine function, one on the cosine function. Within each family the functions can be seen to differ only in their phase shifts.

    a. $y = 2\sin\left(\frac{1}{3}x\right)$ **b.** $y = 5\sin 2x$ **c.** $y = 4\cos 4x$

    d. $y = -3\sin\left(\frac{1}{5}x\right)$ **e.** $y = 1.5\cos\left(\frac{1}{4}x\right)$ **f.** $y = 2 - \cos 2x$

    g. $y = -5\cos 2x$ **h.** $y = 10 + 6\sin 8x$ **i.** $y = -3 + 3\sin 4x$

    j. $y = 3 + 3\cos 4x$ **k.** $y = \sin\left(\frac{4}{2}x\right)$ **l.** $y = 4\cos \pi x$

19. The vertical shift is 59. The amplitude of a sinusoidal model would be one-half the difference of the maximum and minimum values, $(1/2)(71 - 47) = 12$. The period would be (naturally) 24 hours, so the frequency $(2\pi)/24 = \pi/12$. The equation of a rough sinusoidal model might be $59 + 12 \sin\left(\frac{\pi}{12}(t-9)\right)$. One could then tweak the model to get a closer correspondence to the observed data.

21. We assume a model of the form $D + A \sin(B(t - C))$, where we take t to represent the day of the year and assume a year of 365 days. The natural period is 365 days, which gives $B = 2\pi/365$. The average of all measured temperatures is 76°. This gives $D = 76$. The maximum and minimum temperatures are 99° and 53° respectively. Each differs by 23° from the average. Based on this information we take $A = 23$. Determining the phase shift C is more delicate. We can estimate that the daily high temperature was 76 ($\sin(B(t - C)) = 0$) somewhere between $t = 91$ and $t = 105$. Over this period of fourteen days, the daily high temperature rose five degrees from 72° to 77°, so we can assume that the temperature rose one degree every three days. Using this we estimate that the Dallas daily high temperature hit 76° on or about day 88. That is, we take $C = 88$. The formula we have constructed is $76 + 23 \sin\left(\frac{2\pi}{365}(t-88)\right)$. In order to arrive at this formula, we implicitly assumed that we recorded the high and low temperatures of the year on the days sampled. This is unlikely, and the formula can be tweaked experimentally from here.

23. The function $R(t) = 0.85 \sin\left(\frac{2\pi}{5}t\right)$ has a period of 5 seconds, so in 1 minute a person breathes

$$\frac{60}{5} = 12 \text{ seconds}.$$

25. **a.** With t in seconds, the period is 30 seconds.

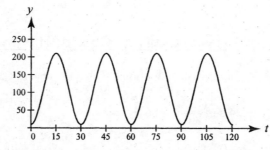

b. Height is given by $110 + 100\sin\left(\frac{\pi}{15}(t-7.5)\right)$.

c. The horizontal distance is given by $x = -100\sin\left(\frac{\pi}{15}t\right)$.

d. The intervals of t for which you are moving forward are $7.5 < t < 22.5$, $37.5 < t < 22.5$, and so on. Assuming that you are facing to your right, you will be moving forward on the top half of the cycle.

e. $x = 100\sin\left(\frac{\pi}{15}t\right)$; for the intervals $-7.5 < t < 7.5$, $22.5 < t < 37.5$, and so on.

f. $(y-110)^2 + x^2 = 10,000$.

27. The average brightness is $\dfrac{4.35+3.65}{2}=4$ and the amplitude of the variation in $\dfrac{4.35-3.65}{2}=0.35$. Using a cosine function and shifting by half a period to put the minimum at $t=0$ we get

$$B(t)=4+0.35\cos\left(\frac{2\pi}{5.4}\left(t-\frac{5.4}{2}\right)\right) \text{ or } B(t)=4+0.35\cos\left(\frac{\pi}{2.7}(t-2.7)\right).$$ Using a sine function, we would get

$$B(t)=4+0.35\sin\left(\frac{\pi}{2.7}(t-1.35)\right).$$

29. a. One possible model based on the scatter plot is that the average global temperature is oscillating above and below the linear trend with an amplitude of about 0.1° and a period of about 60 years. A

corresponding sine function would be $E(t)=0.1\sin\left(\frac{2\pi}{60}(t-45)\right)$, where the phase shift value 45 is

chosen so that the fluctuation E peaks at $t=60$.

b. If we add this to the linear fit, the plot looks like this, over the same range as shown in the scatter plot:

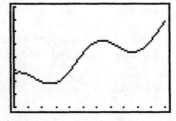

c. For the year 2005 this function predicts an average global temperature of

$0.0042(125)+14.67+E(125)=15.3°.$

Section 7.3   Solving Equations with Sine and Cosine: The Inverse Functions

1. The number of hours of daylight in San Diego is approximated by $H=12+2.4\sin\left(\frac{2\pi}{365}(t-80)\right)$.

For $H=11$ hours: $11=12+2.4\sin\left(\frac{2\pi}{365}(t-80)\right)$

$$-0.4167=\sin\left(\frac{2\pi}{365}(t-80)\right)$$

$$\arcsin(-0.4167)=\arcsin\left[\sin\left(\frac{2\pi}{365}(t-80)\right)\right]$$

$$\arcsin(-0.4167)=\frac{2\pi}{365}(t-80)$$

$$t-80=\frac{\arcsin(-0.4167)\cdot365}{2\pi}$$

$t=55$

Using the symmetry of the sine function, there is a second solution:

$$\pi-\arcsin(-0.4167)=\frac{2\pi}{365}(t-80)$$

$$t-80=\frac{\left[\pi-\arcsin(-0.4167)\right]\cdot365}{2\pi}$$

$$t=287$$

San Diego has 11 hours of daylight on Feb 24 and Oct 14.

For $H = 10$ hours: $10 = 12 + 2.4 \sin\left(\frac{2\pi}{365}(t-80)\right)$

$$-0.8333 = \sin\left(\frac{2\pi}{365}(t-80)\right)$$

$$t - 8 - = \frac{\arcsin(-0.8333)\cdot 365}{2\pi}$$

$$t = 23$$

or

$$t - 80 = \frac{\left[\pi - \arcsin(-0.8333)\right]\cdot 365}{2\pi}$$

$$t = 320$$

San Diego has 10 hours of daylight on Jan 23 and Nov 16.

For $H = 9$ hours: $9 = 12 + 2.4 \sin\left(\frac{2\pi}{365}(t-80)\right)$

$$-1.25 = \sin\left(\frac{2\pi}{365}(t-80)\right)$$

$$\arcsin(-1.25) = \frac{2\pi}{365}(t-80)$$

There is no solution because arcsin (-1.25) is undefined (sin x can never be -1.25).

3. **a.** The temperature reaches a maximum at $69° + 3° = 72°$.

   b. The temperature reaches a minimum at $69° - 3° = 66°$.

   c.
   $$T(t) = 69 + 3\sin\frac{\pi t}{10}$$

   $$70 = 69 + 3\sin\frac{\pi t}{10}$$

   $$\frac{1}{3} = \sin\frac{\pi t}{10}$$

   $$\arcsin\left(\frac{1}{3}\right) = \frac{\pi t}{10}$$

   $$1.08 = t$$

The temperature cycles every 20 minutes, so it reaches 70° at roughly 1 minute and again at 9 minutes into its cycle.

$$T(t) = 69 + 3\sin\frac{\pi t}{10}$$

$$67 = 69 + 3\sin\frac{\pi t}{10}$$

$$-\frac{2}{3} = \sin\frac{\pi t}{10}$$

$$\arcsin\left(-\frac{2}{3}\right) = \frac{\pi t}{10} \quad \text{or} \quad \pi - \arcsin\left(-\frac{2}{3}\right) = \frac{\pi t}{10}$$

$$-2\text{ min }18\text{ sec} = t \qquad\qquad 12\text{ min }18\text{ sec} = t$$

The temperature cycles every 20 minutes, so it reaches 67° at roughly 12 minutes 18 seconds and again at 20 minutes $-$ (2 minutes 18 seconds), or 17 minutes 42 seconds.

5. The model we developed for the tides at the Bay of Fundy was $25 + 25 \sin(2\pi t/11)$.

   a. In our model, low tide corresponds to $25 + 25 \sin(2\pi t/11) = 0$, hence to $t = 8.25$ hours. Solving $25 + 25 \sin(2\pi t/11) = 5$ yields $t = 9.38$ hours as the first solution after 8.25. It takes roughly one hour for the water to rise 5 feet after low tide.

   b. Average tide occurs in our model at $t = 0$, at which point the depth of water is measured as 25 feet. Solving $25 + 25 \sin(2\pi t/11) = 25$, gives $t = 0.352518$. Thus it takes just 21 minutes for the water to rise 5 feet at average tide.

7. Our Fairbanks high temperature function from Problem 15 in Section 7.2 is

$$T(t) = 22 + 42 \sin\left(\frac{2\pi}{365}(t - 131)\right),$$ with t in days from the beginning of the year. T will be 0 when

$$\sin\left(\frac{2\pi}{365}(t-131)\right) = -\frac{22}{42} = -0.524. \text{ Then } \frac{2\pi}{365}(t-131) = \arcsin(-0.524) = -0.552 \text{ and}$$

$$t = \left(-0.552 \cdot \frac{365}{2\pi} + 131\right) = 98.9. \text{ Thus the maximum daily temperature should be } 0° \text{ around the } 99^{\text{th}} \text{ day, or}$$

April 9. There is another solution later in the year when temperatures are on the way back down. Recall that the coldest day of the year is day 40. Our second $0°$ day will be as far *before* the coldest day next year as day 99 is *after* the coldest day this year. Thus another solution is $(365 + 40) - (99 - 40) = 346$ or December 12.

9. The function we constructed in Problem 20 of Section 7.2 is $T(t) = 70.5 + 6.6 \sin(0.5236(t - 5.7))$, with t in months, $t = 0$ corresponding to the beginning of January.

   a. $65 = 70.5 + 6.6 \sin(0.5236(t - 5.7))$

$$\sin(0.5236(t - 5.7)) = \frac{65 - 70.5}{6.6} = -0.8333$$

$$0.5236(t - 5.7) = \arcsin(-0.8333) = -0.9851$$

$$t = \frac{-0.9851}{0.5236} + 5.7 = 3.82$$

This solution is near the end of April. Another solution is located symmetrically to the left of the minimum at 2.7, namely at $2.7 - (3.82 - 2.7) = 1.58$, in the middle of February.

   b. $70 = 70.5 + 6.6 \sin(0.5236(t - 5.7))$

$$\sin(0.5236(t - 5.7)) = \frac{70 - 70.5}{6.6} = -0.0758$$

$$0.5236(t - 5.7) = \arcsin(-0.0758) = -0.0759$$

$$t = \frac{-0.0759}{0.5236} + 5.7 = 5.56$$

This solution is in mid-May. There is another solution located symmetrically on the other side of the maximum at 8.7, namely at $8.7 + (8.7 - 5.56) = 11.84$, near the end of December.

   c. According to the model, the average daytime high does not reach $80°$.

Exercising Your Algebra Skills

1. No solutions: $|\sin x| \leq 1$ for all x.

2. Basic solution: $\theta = \arcsin(0.4) = 0.412$

   General solution: $\theta = 0.412 + 2n\pi$ and $\theta = (\pi - 0.412) + 2n\pi = 2.730 + 2n\pi$

3. No solutions: $|\sin x| \le 1$ for all x.

4. Basic solution: $\theta = \arcsin(0.75) = 0.848$
    General solution: $\theta = 0.848 + 2n\pi$ and $\theta = (\pi - 0.848) + 2n\pi = 2.294 + 2n\pi$

5. Basic solution: $\theta = \arcsin(-0.75) = -0.848$
    General solution: $\theta = -0.848 + 2n\pi$ and $\theta = (\pi - (-0.848)) + 2n\pi = 3.990 + 2n\pi$

6. Basic solution: $2x = \arcsin(0.6) = 0.644$
    General solution: $2x = 0.644 = 2n\pi$ and $2x = (\pi - 0.644) + 2n\pi = 2.498 + 2n\pi$
    Thus, $x = 0.322 + n\pi$ and $x = 1.249 + n\pi$

7. Basic solution: $2x = \arccos(-0.6) = 2.214$
    General solution: $2x = 2.214 + 2n\pi$ and $2x = (2\pi - 2.214) + 2n\pi = 4.069 + 2n\pi$
    Thus, $x = 1.107 + n\pi$ and $x = 2.034 + n\pi$

8. Basic solution: $x = \arccos\left(\dfrac{2}{3}\right) = 0.841$

    General solution: $x = 0.841 + 2n\pi$ and $x = (2\pi - 0.841) + 2n\pi = 5.442 + 2n\pi$

9. Using a double-angle formula from Section 8.1, we can write this equation as
    $5(2\sin x \cos x) = 3\cos x$ or $10\sin x \cos x = 3\cos x$. Now either $\cos x = 0$, which happens when

    $x = \dfrac{\pi}{2} + n\pi$, or we can divide through by $\cos x$ to get $\sin(x) = 0.3$. This equation has basic solution

    $x = \arcsin(0.3) = 0.305$ and general solution $x = 0.305 + 2n\pi$ and $x = (\pi - 0.305) + 2n\pi = 2.837 + 2n\pi$.

    Thus the complete solution is $x = \dfrac{\pi}{2} + n\pi, \ 0.305 + 2n\pi, \ 2.837 + 2n\pi$.

Section 7.4   The Tangent Function

1. **a.** $y = \tan(2x)$

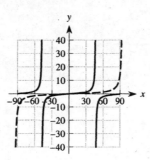

b. $y = \tan\left(\dfrac{1}{2}x\right)$

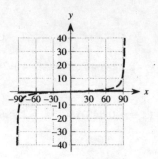

c. $y = 3\tan x$

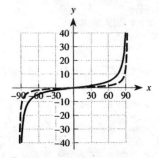

d. $y = -2\tan x$

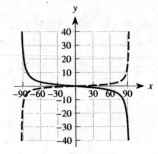

e. $y = \tan(x - 30)$

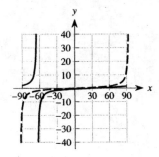

f. $y = -10 + \tan(x)$

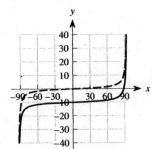

3. The standard tangent graph has asymptotes at $\theta = \pm\pi/2$ and the table indicates asymptotes at $\pm\pi/3$, so we need to compress the graph horizontally by a factor of 3/2. The equation is $y = \tan\left(\frac{3}{2}x\right)$. This checks with the other values in the table.

5.

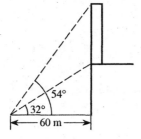

The height b of the building is given by $\tan 32° = \dfrac{b}{60}$. Thus $b = 60\tan 32° = 37.5$ meters. The distance d from the top of the smokestack to the ground is $d = 60\tan 54° = 82.6$ meters. The height of the smokestack is $d - g = 45.1$ meters.

7. The slope m of a line is computed as rise/run. However, for a line through the origin, the line will pass through the point $(\cos\theta,\ \sin\theta)$, where θ is the angle formed by the given line and the x-axis. Using this point and the origin to compute the slope we get $m = \sin\theta/\cos\theta = \tan\theta$. The angles formed by the lines $y = x$, $y = 2x$, $y = 3x$, $y = 4x$, are $\pi/4$, arctan(2), arctan(3), and arctan(4), respectively. The line whose equation is $y = mx + b$ may be viewed as the translation by b units in the y-direction of the line $y = mx$. Since the angle formed by a line and the x-axis is not changed by a translation of the line, we may always view the slope of a line as the tangent of the angle formed by the line and the x-axis.

9.

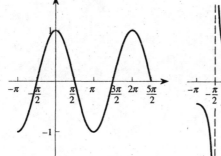

11.

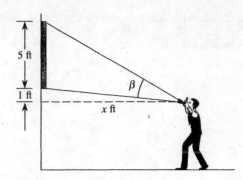

a. The angle from the horizontal to the top of the painting is $\arctan\left(\dfrac{6}{x}\right)$ and the angle to the bottom is

$\arctan\left(\dfrac{1}{x}\right)$, so $\beta = \arctan\left(\dfrac{6}{x}\right) - \arctan\left(\dfrac{1}{x}\right)$.

b.

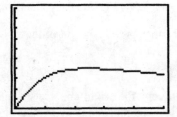

The graph indicates a maximum angle at a distance from the wall of about 2.4 or 2.5 feet. The maximum angle is close to 45°.

13. **a.**

v	0	0.5c	0.9c	0.95c	0.99c	0.999c
M	1	1.155	2.294	3.203	7.089	22.366

b.

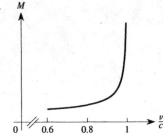

c. The actual function giving mass can be reasonably approximated by $M = 1 + \dfrac{\tan\left(\dfrac{\pi}{2}v\right)}{6}$.

Exercising Your Algebra Skills

1. Basic solution: $\theta = \arctan\left(\dfrac{5}{4}\right) = 0.896$

General solution: $\theta = 0.896 + n\pi$

2. Basic solution: $\theta = \arctan\left(\dfrac{4}{5}\right) = 0.675$

   General solution: $\theta = 0.675 + n\pi$

3. Basic solution: $\theta = \arctan(-1) = -\dfrac{\pi}{4}$

   General solution: $\theta = -\dfrac{\pi}{4} + n\pi$, which we can also write as $\theta = \dfrac{3\pi}{4} + n\pi$.

4. Basic solution: $\theta = \arctan(2) = 1.107$

   General solution: $\theta = 1.107 + n\pi$

5. Basic solution: $\theta = \arctan\left(\dfrac{4}{5}\right) = 0.675$

   General solution: $\theta = 0.675 + n\pi$

6. Basic solution: $\theta = \arctan\left(\dfrac{5}{4}\right) = 0.896$

   General solution: $\theta = 0.896 + n\pi$

7. $\sin\theta + \cos\theta = 0 \Rightarrow \sin\theta = -\cos\theta \Rightarrow \theta = \arctan(-1) = -\dfrac{\pi}{4}$

   General solution: $\theta = -\dfrac{\pi}{4} + n\pi$, which can be written as $\theta = \dfrac{3\pi}{4} + n\pi$.

8. $4\sin\theta = 3\cos\theta \Rightarrow \theta = \arctan\left(\dfrac{3}{4}\right) = 0.644$

   General solution: $\theta = 0.644 + n\pi$

Chapter 7 Review Problems

1. The answers are the same because $\cos x = \sin\left(x + \dfrac{\pi}{2}\right)$ and $5.70 + \dfrac{\pi}{2} \approx 7.2708$.

2. a. $\theta = 120°, 420°, 480°, \dots$ $\theta = -240°, -300°, -600°, -660°, \dots$
   b. $t = 2\pi/3, 7\pi/3, 8\pi/3, \dots$ $t = -4\pi/3, -5\pi/3, -10\pi/3, \dots$

3. a. $\theta = 315°, 405°, 675°\dots$ $\theta = -45°, -315°, -405°, -675°\dots$
   b. $\theta = 7\pi/4, 9\pi/4, 15\pi/4\dots$ $\theta = -\pi/4, -7\pi/4, -9\pi/4, -15\pi/4\dots$

	vertical shift	amplitude	frequency	period	phase shift
4.	325	10	$2\pi/9$	9	0
5.	63	3	$2\pi/25$	25	0
6.	71	2	$2\pi/15$	15	0
7.	80	13	$\pi/12$	24	15
8.	38	8	$\pi/12$	24	5
9.	100	25	$\pi/36$	72	0
10.	100	25	$2\pi/97$	97	0
11.	145	40	$2\pi/83$	83	0

12–19. Answers will vary.

20. The side view of the swing is shown. For any θ between $0°$ and $60°$, or 0 and $\frac{\pi}{3}$ radians, the hypotenuse is 8 ft because that is the length of the chain. At the maximum, when $\theta = \frac{\pi}{3}$, the vertical side of the triangle shown is $8\cos\frac{\pi}{3} = 4$ ft and the base of the triangle is 693 ft.

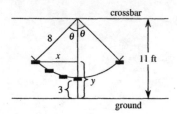

a. For the vertical motion, the height of the seat oscillates between 3 feet above ground level and $11 - 4 = 7$ feet, so it is centered at a height of 5 feet above ground level. Since this happens over a 3-second interval, one sinusoidal model for the height is $y = 5 + 2\cos\left(\frac{2\pi}{3}t\right)$, where t is in radians, which starts at the maximum height of 7 feet when $t = 0$.

b. For the horizontal motion, the distance left and right of center oscillates between -6.93 and 6.93 feet, with an average value of 0. However, during 3 seconds this motion completes only half a cycle (say from the far right to the far left). Thus, one model is $x = 6.93\cos\left(\frac{2\pi}{6}t\right)$, which starts at the maximum horizontal distance that corresponds to the maximum height of $y = 7$ at time $t = 0$. The 6 in the denominator gives the required period of 6 seconds.

21. If the bungee cord didn't contract, the jumper would simply oscillate at a maximum of 40 feet above and below the 160-feet level. So a simple sinusoidal model would be $D = 160 + 40\cos\frac{2\pi}{6}t$, which starts at a distance $d = 200$ feet below the bridge at time $t = 0$. To account for the decaying oscillation, which dies out over the first 10 cycles, we use a function of the form $D = 160 + 40a^t\cos\frac{2\pi}{6}t$ for some $a < 1$. Suppose we interpret the information that the oscillation dies out after 10 cycles of length 6 seconds to mean that the distance after the tenth oscillation is within 0.5 foot of the 160-feet level mark since theoretically the jumper never comes to a complete stop. So, when $t = 60$, we need to solve the following equation:

$$160 + 40a^{60}\cos\frac{2\pi}{6}\cdot 60 = 160.5$$
$$160 + 40a^{60}\cos 20\pi = 160.5$$
$$160 + 40a^{60} = 160.5$$
$$40a^{60} = 0.5$$
$$a^{60} = 0.0125$$
$$a = (0.0125)^{1/60} = 0.92957$$

Thus, our model is $D = 160 + 40(0.92957)^t\cos\frac{\pi}{3}t$.

22. **a.** $H = 1.5\sin\left(\dfrac{2\pi}{500}t\right)$

b. $H = 1.5 + 1.5\sin\left(\dfrac{2\pi}{500}t\right)$

23. The pedals start at a height of $4 + 10/2 = 9$ inches and the amplitude is 5 inches. The function might be $H = 9 + 5\sin\left(\dfrac{2\pi}{3}t\right)$ if the child is going up at time 0.

24. **a.** If we add a uniform horizontal speed of 10 inches per second and take the midline as the average position of the child's feet, the graph of $y = 5\sin\left(\dfrac{2\pi}{3}\cdot\dfrac{x}{10}\right)$ might indicate the height y as a function of the distance x traveled, so that a plot of y versus x gives a picture of the path of the child's feet through space.

b. The graph of the function over 10 seconds looks like this:

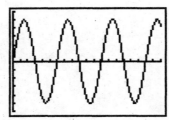

The graph makes it clear that our model is suspect: there is no indication in the graph that conditions are different when the stick is in contact with the ground.

c.

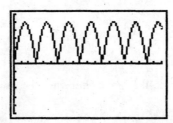

This graph looks a bit more realistic, but it is still wrong. Unless the stick slips on the floor, the child's feet can't move forward much while the stick is in contact with the floor (they can move forward a bit as the stick rotates around the contact point) but this graph still has uniform forward motion in the x-direction.

25. **a.** $s = 4000\cdot\dfrac{\pi}{4} = 3142$ miles

b. $s = 4000\cdot\dfrac{\pi}{3} = 4189$ miles

c. $s = 4000\cdot\dfrac{5\pi}{6} = 10{,}472$ miles

d. $s = 4000\cdot15°\cdot\dfrac{\pi}{180°} = 1047$ miles

26. **a.** $\dfrac{180 \text{ rev}}{\text{min}}\cdot\dfrac{2\pi \text{ radians}}{\text{rev}} = 360\pi\ \dfrac{\text{radians}}{\text{min}}$

b. $\dfrac{2 \text{ ft}}{\text{radian}}\cdot\dfrac{360\pi \text{ radians}}{\text{min}}\cdot 1\text{min} = 720\pi \text{ feet}$

27. Numerical answers will vary depending on the diameter D. With D measured in inches, the tire makes $\dfrac{5280\cdot 12}{\pi D}$ revolutions per minute.

28.

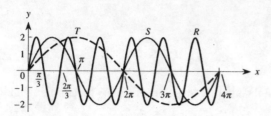

29. a. Frequency $\frac{3}{4}$, period $\frac{8\pi}{3}$, amplitude 2, and phase shift 0.

  b. Frequency $\frac{3}{4}$, period $\frac{8\pi}{3}$, amplitude 2, and phase shift $-\frac{4\pi}{3}$.

  c. Frequency π, period 2, amplitude 2, and phase shift $-\frac{3}{4\pi}$.

  d. Frequency $\frac{3\pi}{4}$, period $\frac{8}{3}$, amplitude 2, and phase shift $\frac{1}{\pi}$.

30. a. $x = \pm\frac{4\pi}{9} + \frac{8\pi\, n}{3}$ **b.** $x = \pm\frac{4\pi}{9} + \frac{8\pi\, n}{3}$

  c. $x = \frac{2}{3} - \frac{3}{4\pi} + 2n$ or $x = \frac{4}{3} - \frac{3}{4\pi} + 2n$ **d.** $x = \frac{8}{9} + \frac{1}{\pi} + \frac{8n}{3}$ or $x = \frac{16}{9} + \frac{1}{\pi} + \frac{8n}{3}$

31. a.

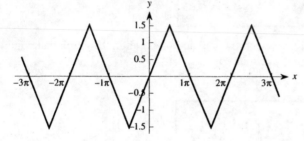

The graph shows that arcsin(sin x) is not the same as the identity function x: Over the interval $-\frac{\pi}{2} \le x \le \frac{\pi}{2}$, the graph *does* look like $y = x$, but since the values of the inverse sine function are always between $-\frac{\pi}{2}$ and $\frac{\pi}{2}$, inputs to the since function outside this range will get moved back into this range by the function arcsin(sin x).

  b.

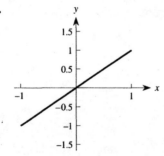

32. a. $\theta = \arctan\frac{3}{2} + n\pi = 0.98 + n\pi$ **b.** $\theta = \arctan 2 + n\pi = 1.11 + n\pi$

  c. $\theta = \arctan 7 + n\pi = 1.43 + n\pi$

33. a. $x = 4.45522$ **b.** $x = 0.47943$

Chapter 8   More About the Trigonometric Functions

Section 8.1   Relationships Among Trigonometric Functions

1. The amplitude of the graph of $y = \cos 3x$ is 1 while the amplitude of $y = 3\cos x$ is 3. The frequency of $y = \cos 3x$ is 3 while the frequency of $y = 3\cos x$ is 1. Thus $\cos 3x \neq 3\cos x$.

3. Evaluating $\sin^3 x + \cos^3 x$ at $x = 1$, we get 0.75355, not 1. So, $\sin^3 x + \cos^3 x = 1$ is not an identity.

5. $\dfrac{\sin 2x}{\sin x} = \dfrac{2\sin x \cos x}{\sin x} = 2\cos x$

7. Taking $x = \frac{\pi}{3}$,, we have $\sin\left(3 \cdot \frac{\pi}{3}\right) = \sin \pi = 0$, whereas $3\sin\frac{\pi}{3} = \frac{3\sqrt{3}}{2}$. Since the two sides of the equation fail to agree at $x = \frac{\pi}{3}$, the equation $\sin 3x = 3\sin x$ is not an identity.

9. $\begin{aligned}\dfrac{1-\cos\alpha}{\sin\alpha} &= \dfrac{1-\cos\alpha}{\sin\alpha} \cdot \dfrac{1+\cos\alpha}{1+\cos\alpha}\\ &= \dfrac{1-\cos^2\alpha}{\sin\alpha(1+\cos\alpha)}\\ &= \dfrac{\sin^2\alpha}{\sin\alpha(1+\cos\alpha)}\\ &= \dfrac{\sin\alpha}{1+\cos\alpha}\end{aligned}$

11. $\begin{aligned}\sin^2 3x + \cos 6x &= \sin^2 3x + \cos(2 \cdot 3x)\\ &= \sin^2 3x + \cos^2 3x - \sin^2 3x\\ &= \cos^2 3x\end{aligned}$

13. Taking $x = 0$, we find that $\sin(\cos 0) = \sin 1 = 0.8415$ and $\cos(\sin 0) = \cos 0 = 1$. Since the two sides of the equation fail to agree at $x = 0$, the equation $\sin(\cos x) = \cos(\sin x)$ is not an identity.

15. $\begin{aligned}\cos 3x &= \cos(x + 2x)\\ &= \cos x \cos 2x - \sin x \sin 2x\\ &= \cos x(\cos^2 x - \sin^2 x) - \sin x(2\sin x \cos x)\\ &= \cos^3 x - \cos x \sin^2 x - 2\cos x \sin^2 x\\ &= \cos^3 x - 3\cos x \sin^2 x\\ &= \cos^3 x - 3\cos x(1-\cos^2 x)\\ &= 4\cos^3 x - 3\cos x\end{aligned}$

17. $\cos 5x = \cos(x + 4x)$

$= \cos x \cos 4x - \sin x \sin 4x$

$= \cos x (8\cos^4 x - 8\cos^2 x + 1) - \sin x (2\sin 2x \cos 2x)$

$= \cos x (8\cos^4 x - 8\cos^2 x + 1) - \sin x [2(2\sin x \cos x)(\cos^2 x - \sin^2 x)]$

$= 8\cos^5 x - 8\cos^3 x + \cos x - \sin x (4\sin x \cos^3 x - 4\sin^3 x \cos x)$

$= 8\cos^5 x - 8\cos^3 x + \cos x - \sin x [4\sin x \cos x (1 - \sin^2 x) - 4\sin x \cos x (1 - \cos^2 x)]$

$= 8\cos^5 x - 8\cos^3 x + \cos x - 4\sin^2 x \cos x (1 - \sin^2 x - 1 + \cos^2 x)$

$= 8\cos^5 x - 8\cos^3 x + \cos x - 4\sin^2 x \cos x (2\cos^2 x - 1)$

$= 8\cos^5 x - 8\cos^3 x + \cos x - 8\sin^2 x \cos^3 x + 4\sin^2 x \cos x$

$= 8\cos^5 x - 8\cos^3 x + \cos x - 8(1 - \cos^2 x)\cos^3 x + 4(1 - \cos^2 x)\cos x$

$= 8\cos^5 x - 8\cos^3 x + \cos x - 8\cos^3 x + 8\cos^5 x + 4\cos x - 4\cos^3 x$

$= 16\cos^5 x - 20\cos^3 x + 5\cos x$

19.   We substitute $y = x$ into Equation (7):

$$\cos(x + y) = \cos x \cos y - \sin x \sin y$$
$$\cos(x + x) = \cos x \cos x - \sin x \sin x$$
$$\cos(2x) = \cos^2 x - \sin^2 x$$

21. $\sin^2 x \cos^2 x = \frac{1}{2}(1 - \cos 2x) \cdot \frac{1}{2}(1 + \cos 2x)$

$= \frac{1}{4}(1 - \cos^2 2x)$

$= \frac{1}{4}(1 - \frac{1}{2}[\cos(2 \cdot 2x)])$

$= \frac{1}{4}(\frac{1}{2} - \frac{1}{2}\cos 4x)$

$= \frac{1}{8} - \frac{1}{8}\cos 4x$

23.   **a.**

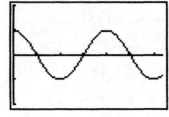

$y = \sin(x + \pi) = -\sin x$

$\sin(x + \pi) = \sin x \cos \pi + \cos x \sin \pi = \sin x \cdot (-1) + \cos x \cdot 0 = -\sin x$

b.

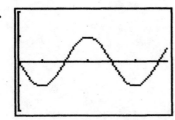

$y = \cos(x + \pi/2) = -\sin x$

$\cos\left(x + \frac{\pi}{2}\right) = \cos x \cos \frac{\pi}{2} - \sin x \sin \frac{\pi}{2} = \cos x \cdot 0 - \sin x \cdot 1 = -\sin x$

25.

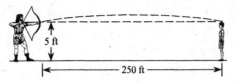

$$5 \text{ ft}$$

$$\longleftarrow \text{—— } 250 \text{ ft } \text{——} \longrightarrow$$

Using the formula $R = \dfrac{2v_0{}^2 \sin 2\theta}{g}$, we have

$$250 = \frac{2(180)^2 \sin 2\theta}{32} \Rightarrow \sin 2\theta = \frac{10}{81} = 0.1235 \Rightarrow$$

$$2\theta = \arcsin(0.1235) \Rightarrow 2\theta = 7.09° \quad \text{or} \quad 2\theta = 180° - 7.09° = 172.91°$$

$$\theta = 3.55° \quad \text{or} \quad \theta = 86.45°$$

The nearly vertical trajectory should be rejected as unsafe.

27. **a.** To shift the cosine graph $90°$ to the right to match the corresponding sine graph, we subtract $90°$ from the argument, giving $y = 5\cos t(t - 126.8°)$.

b. Using $C\cos(t - D) = C(\cos t \cos D + \sin t \sin D)$, we let $C = 5$, $\sin D = \frac{4}{5}$, and $\cos D = -\frac{3}{5}$. Then D is an angle in the second quadrant with tangent equal to $-\frac{4}{3}$. Since $\arctan\left(-\frac{4}{3}\right) = -53.13°$,

$$D = 180° - 53.13 = 126.87°, \text{ and our formula is } y = 5\cos(t - 126.87°).$$

c. The results in parts (a) and (b) agree.

29. **a.** The vertical shift is $\frac{3}{4}$, the amplitude is $\frac{1}{4}$, and the frequency is 4, so a reasonable candidate is

$$y = \frac{3}{4} + \frac{1}{4}\cos 4x.$$

b. The two graphs appear identical.

c. Squaring both sides of $\sin^2 x = \frac{1}{2}(1 - \cos 2x)$ and $\cos^2 x = \frac{1}{2}(1 + \cos 2x)$ and adding, we get:

$$\sin^4 x + \cos^4 x = \frac{1}{2}(1 + \cos^2 2x) \tag{1}$$

Using the second half-angle identity with $2x$ in place of x gives $\cos^2 2x = \frac{1}{2}(1 + \cos 4x)$. Substituting for $\cos^2 2x$ into the right side of Equation (1) above, we get:

$$\sin^4 x + \cos^4 x = \frac{1}{2}\left[1 + \frac{1}{2}(1 + \cos 4x)\right] = \frac{3}{4} + \frac{1}{4}\cos 4x$$

31. The graph in Problem 1(a) of Section 7.2 on depicts a function that is even.
The graph in Problem 1(b) of Section 7.2 depicts a function that is odd.
The graph in Problem 1(c) of Section 7.2 depicts a function that is even.
The graph in Problem 1(d) of Section 7.2 depicts a function that is neither even nor odd.
The graph in Problem 1(e) of Section 7.2 depicts a function that is neither even nor odd.
The graph in Problem 1(f) of Section 7.2 depicts an even function.
The graph in Problem 1(g) of Section 7.2 depicts a function that is neither even nor odd.
The graph in Problem 1(h) of Section 7.2 depicts a constant function. Constant functions are always even.
The graph in Problem 1(i) of Section 7.2 depicts a function that is neither even nor odd.
The graph in Problem 1(j) of Section 7.2 depicts a function that is neither even nor odd.

33. Substitute $x = \pi/6$ into the formula of Problem 32(b):

for $n = 2$ $1 + \sin\frac{\pi}{6} + \sin^2\frac{\pi}{6} = \dfrac{1 - \sin^3\frac{\pi}{6}}{1 - \sin\frac{\pi}{6}} = \dfrac{1 - \left(\frac{1}{2}\right)^3}{1 - \frac{1}{2}} = \dfrac{\frac{7}{8}}{\frac{1}{2}} = \dfrac{7}{4} = 1.75$

for $n = 3$ $1 + \sin\frac{\pi}{6} + \sin^2\frac{\pi}{6} + \sin^3\frac{\pi}{6} = \dfrac{1 - \sin^4\frac{\pi}{6}}{1 - \sin\frac{\pi}{6}} = \dfrac{1 - \left(\frac{1}{2}\right)^4}{1 - \frac{1}{2}} = \dfrac{\frac{15}{16}}{\frac{1}{2}} = \dfrac{15}{8} = 1.875$

for $n = 4$ $1 + \sin\frac{\pi}{6} + \sin^2\frac{\pi}{6} + \sin^3\frac{\pi}{6} + \sin^4\frac{\pi}{6} = \dfrac{1 - \sin^5\frac{\pi}{6}}{1 - \sin\frac{\pi}{6}} = \dfrac{1 - \left(\frac{1}{2}\right)^5}{1 - \frac{1}{2}} = \dfrac{\frac{31}{32}}{\frac{1}{2}} = \dfrac{31}{16} = 1.938$

Using the infinite sum formula, we know that $1 + \sin\frac{\pi}{6} + \sin^2\frac{\pi}{6} + \cdots = \dfrac{1}{1 - \sin\frac{\pi}{6}} = 2$. Looking at the different partial sums, we see that it takes a partial sum of 12 terms to achieve three-decimal place accuracy. For $n = 11$, the partial sum equals 1.9995. Since $\sin\frac{\pi}{6} = 0.5$ and $\sin\frac{\pi}{3} = 0.8660$, the powers of $\sin\frac{\pi}{3}$ get smaller much faster than the powers of $\sin\frac{\pi}{6}$. So we would expect that you will need more terms to achieve the same three-place accuracy. The infinite sum when $x = \frac{\pi}{3}$ is 7.464. Calculating the partial sum with 12 terms, $1 + \sin\frac{\pi}{3} + \sin^2\frac{\pi}{3} + \cdots + \sin^{11}\frac{\pi}{3}$, is only 3.463256. So you need more than 12 terms to achieve the three-decimal place accuracy of 7.464.

35. $\dfrac{1}{\tan x} - \tan x = \dfrac{1}{\frac{\sin x}{\cos x}} - \dfrac{\sin x}{\cos x}$

$= \dfrac{\cos x}{\sin x} - \dfrac{\sin x}{\cos x}$

$= \dfrac{\cos^2 x - \sin^2 x}{\sin x \cos x}$

$= 2 \cdot \dfrac{\cos 2x}{2\sin x \cos x}$

$= 2 \cdot \dfrac{\cos 2x}{\sin 2x}$

$= 2 \cdot \dfrac{1}{\frac{\sin 2x}{\cos 2x}}$

$= \dfrac{2}{\tan 2x}$

37. **a.** Using the double-angle formulas for sine and cosine, we have:

$$\tan 2x = \frac{\sin 2x}{\cos 2x}$$

$$= \frac{2\sin x \cos x}{\cos^2 x - \sin^2 x}$$

$$= \frac{2\sin x \cos x}{\cos^2 x - \sin^2 x} \cdot \frac{\dfrac{1}{\cos^2 x}}{\dfrac{1}{\cos^2 x}}$$

$$= \frac{2\dfrac{\sin x}{\cos x}}{1 - \left(\dfrac{\sin x}{\cos x}\right)^2}$$

$$= \frac{2\tan x}{1 - \tan^2 x}$$

b. Using the addition formula for sine and cosine, we have:

$$\tan(x + y) = \frac{\sin(x + y)}{\cos(x + y)}$$

$$= \frac{\sin x \cos y + \cos x \sin y}{\cos x \cos y - \sin x \sin y}$$

$$= \frac{\sin x \cos y + \cos x \sin y}{\cos x \cos y - \sin x \sin y} \cdot \frac{\dfrac{1}{\cos x \cos y}}{\dfrac{1}{\cos x \cos y}}$$

$$= \frac{\dfrac{\sin x}{\cos x} + \dfrac{\sin y}{\cos y}}{1 - \dfrac{\sin x}{\cos x} \cdot \dfrac{\sin y}{\cos y}}$$

$$= \frac{\tan x + \tan y}{1 - \tan x \tan y}$$

39. Taking $\theta = 1$ the equation gives $\tan 2 = 2 \tan 1$, or $-2.18504 = 3.11482$. Since this is a false statement, the equation is not an identity.

41. $$\tan 2x = \frac{\sin 2x}{\cos 2x}$$

$$= \frac{2\sin x \cos x}{\cos^2 x - \sin^2 x}$$

$$= \frac{2\sin x \cos x}{\cos^2 x - \sin^2 x} \cdot \frac{\dfrac{1}{\cos^2 x}}{\dfrac{1}{\cos^2 x}}$$

$$= \frac{2\dfrac{\sin x}{\cos x}}{1 - \left(\dfrac{\sin x}{\cos x}\right)^2}$$

$$= \frac{2\tan x}{1 - \tan^2 x}$$

43. Letting $x = 0$ in $\tan^2 x = 1 + 2\tan x$ gives $\tan^2 0 = 1 + 2\tan 0$, or $0 = 1 + 0$. Since this $0 = 1$ is a false statement, the equation is not an identity.

45. Substituting $x = 1$ into the equation $\tan(\sin x) = \tan x \sin x$, we get $1.11894 = 1.31051$, we find that the equation is not an identity.

47. The expression $\cos(\tan x)$ means the cosine *of* the value $\tan x$. That is, the expression involves the composition of two functions, not the product. So the expression $\cos(\tan x)$ does not mean $\cos x \cdot \tan x$.

Therefore, the expression $\cos\left(\dfrac{\sin x}{\cos x}\right)$ does not mean $\cos \cdot \dfrac{\sin x}{\cos x}$, and so we cannot cancel "cos."

Section 8.2   Approximating Sine and Cosine with Polynomials

1. With $T_2(x) = 1 - \dfrac{x^2}{2!}$, we have:

$T_2(0) = 1$	$\cos 0 = 1$	$T_2(0.1) = 0.995$	$\cos 0.1 = 0.995004$
$T_2(0.2) = 0.98$	$\cos 0.2 = 0.980067$	$T_2(0.3) = 0.995$	$\cos 0.3 = 0.995336$
$T_2(0.4) = 0.92$	$\cos 0.4 = 0.921061$	$T_2(0.5) = 0.875$	$\cos 0.5 = 0.877583$
$T_2(0.6) = 0.82$	$\cos 0.6 = 0.825336$	$T_2(0.1) = 0.995$	$\cos 0.1 = 0.995004$

3.

x	$T_3(x)$	$\sin x$	$T_3(x) - \sin x$
0	0	0	0
0.1	0.09983	0.09983	-8×10^{-8}
0.2	0.19867	0.19867	-3×10^{-6}
0.3	0.29550	0.29553	-2×10^{-5}
0.4	0.38933	0.38942	-9×10^{-5}
0.5	0.47917	0.47943	-3×10^{-4}
0.6	0.56400	0.56464	-6×10^{-4}

5.

x	$-\dfrac{4\pi}{25}$	$-\dfrac{3\pi}{25}$	$-\dfrac{2\pi}{25}$	$-\dfrac{\pi}{25}$	0	$\dfrac{\pi}{25}$	$\dfrac{2\pi}{25}$	$\dfrac{3\pi}{25}$	$\dfrac{4\pi}{25}$
$\sin x$	-0.4818	-0.3681	-0.2487	-0.1253	0	0.1253	0.2487	0.3681	0.4818

The cubic regression equation is : $\sin x \approx -0.1640x^3 - (2.851 \times 10^{-14})x^2 + 0.998x - (2.886 \times 10^{-15})$; it comes very close to the cubic approximation $\sin x \approx x - x^3/6$.

7. If $T_5(x) = x - \dfrac{x^3}{6} + \dfrac{x^5}{120}$ is a good approximation to $\sin x$, then

$$T_5(x^2) = -x - \frac{x^6}{6} + \frac{x^{10}}{120}$$

should be a good approximation to $g(x) = \sin x^2$. In fact, as the graph on the next page shows, there is a good match for x between -0.5π and 0.5π.

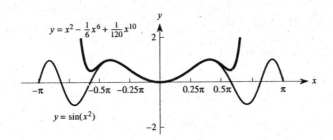

9. $\left(x - \dfrac{x^3}{6}\right)^2 + \left(1 - \dfrac{x^2}{2} + \dfrac{x^4}{24}\right)^2 = 1 - \dfrac{1}{72}x^6 + \dfrac{1}{576}x^8$

This result will get closer and closer to the value 1 when $x = 0$ if you use higher degree approximations.

11. Substituting $\theta = x/2$ into $\cos 2\theta = 2\cos^2\theta - 1$, we get $\cos x = 2\cos^2(x/2) - 1$. Substituting the approximation $\cos(x/2) = 1 - \frac{1}{2}(x/2)^2$, we get the equation:

$$\cos x = 2\left[\left(1 - \frac{1}{2}x^2\right)^2\right]^2 - 1 = 2\left(1 - \frac{x^2}{4} + \frac{x^4}{64}\right) - 1 = 1 - \frac{x^2}{2} + \frac{x^4}{32}$$

13. Approximating $\sin x$ by its linear polynomial in $\dfrac{\sin(x + \Delta x) - \sin x}{\Delta x}$ gives the value of 1. Substituting the cubic Taylor approximation of sine gives:

$$\frac{\sin(x + \Delta x) - \sin x}{\Delta x} \approx \frac{(x + \Delta x) - (x + \Delta x)^3/6 - (x - x^3/6)}{\Delta x}$$

$$= \frac{(x + \Delta x) - (x^3 + 3x^2\Delta x + 3x(\Delta x)^2 + (\Delta x)^3)/6 - x + x^3/6}{\Delta x}$$

$$= \frac{\Delta x - x^2\Delta x/2 - x(\Delta x)^2/2 - (\Delta x)^3/6}{\Delta x}$$

$$= 1 - \frac{x^2}{2} - \frac{x\Delta x}{2} - \frac{(\Delta x)^2}{6}$$

Setting $\Delta x = 0$, we get $1 - x^2/2$. This suggests that for small values of Δx, $\left[\sin(x + \Delta x) - \sin(x)\right]/\Delta x$ is well approximated by $1 - x^2/2$. (Curiously, $1 - x^2/2$ is the quadratic approximation to $\cos x$.)

Section 8.3   Properties of Complex Numbers

1. $\|z\| = \sqrt{4^2 + (-3)^2} = 5$, $\theta = -\arctan(3/4) \approx -0.64350$

3. $\|z\| = \sqrt{12^2 + (-5)^2} = 13$, $\theta = -\arctan(5/12) \approx -0.39479$

5. $\|z\| = \sqrt{64^2 + (-36)^2} = \sqrt{5392} \approx 73.43024$, $\theta = -\arctan(9/16) \approx -0.51239$

7. $\|z\| = \sqrt{(-5)^2 + 7^2} = \sqrt{74} \approx 8.60233$, $\theta = \pi - \arctan(7/5) \approx 2.19105$

9. $\|z\| = \sqrt{(-8)^2 + (-\sqrt{3})^2} = \sqrt{67} \approx 8.18535$, $\theta = \pi + \arctan\left(\sqrt{3}/8\right) \approx 3.35481$

11. $13\left(\cos(1.17601) + i\sin(1.17601)\right)$

13. $25\left(\cos(2.21430) + i\sin(2.21430)\right)$

15. $\sqrt{73}\left(\cos(-0.35877) + i\sin(-0.35877)\right)$

17. $\sqrt{17}\left(\cos(0.75597) + i\sin(0.75597)\right)$

19. $z^2 = (4 - 3i)\bullet(4 - 3i) = 16 - 12i - 12i + 9i^2 = 16 - 24i - 9 = 7 - 24i.$

21. $z^2 = (12 - 5i)\bullet(12 - 5i) = 144 - 60i - 60i + 25i^2 = 144 - 120i - 25 = 119 - 120i.$

For Problems 23 through 31, all equalities are approximate since we are using approximate values for the angles associated with each z.

23. $z^2 = \left(5(\cos(-0.64350) + i\sin(-0.64350))\right)^2$

       $= 25(\cos((2)(-0.64350)) + i\sin((2)(-0.64350)))$

       $= 25(\cos(-1.28700) + i\sin(-1.28700))$

       $= 7.00005 - 23.99998i$

25. $z^2 = \left(13(\cos(-0.39479) + i\sin(-0.39479))\right)^2$

       $= 169(\cos((2)(-0.39479)) + i\sin((2)(-0.39479)))$

       $= 169(\cos(-0.78950) + i\sin(-0.78950))$

       $= 119.00987 - 119.99021i$

27. $z^2 = \left(\sqrt{5392}(\cos(-0.51239) + i\sin(-0.51239))\right)^2$

       $= 5392(\cos((2)(-0.51239)) + i\sin((2)(-0.51239)))$

       $= 5392(\cos(-1.02478) + i\sin(-1.02478))$

       $= 2799.99503 - 4608.00302i$

29. $z^2 = \left(\sqrt{74}(\cos(2.19105) + i\sin(2.19105))\right)^2$

       $= 74(\cos((2)(2.19105)) + i\sin((2)(2.19105)))$

       $= 74(\cos(4.38210) + i\sin(4.38210))$

       $= -23.99941 - 70.00020i$

31. $z^2 = \left(\sqrt{67}(\cos(3.35481) + i\sin(3.35481))\right)^2$

       $= 67(\cos((2)(3.35481)) + i\sin((2)(3.35481)))$

       $= 67(\cos(6.70962) + i\sin(6.70962))$

       $= 60.9999 + 27.71304i$

We use the binomial identity $(x + y)^3 = x^3 + 3x^2y + 3xy^2 + y^3$ in the following four problems.

33. $z^3 = (5 + 12i)^3 = 125 + (3)(25)(12i) + (3)(5)(144i^2) + 1728i^3$

       $= 125 + 900i - 2160 - 1728i = -2035 - 828i$

35. $z^3 = (-15 + 20i)^3 = -3375 + (3)(225)(20i) + (3)(-15)(400i^2) + 8000i^3$

       $= -3375 + 13500i + 18000 - 8000i = 14{,}625 + 5500i$

37. $z^3 = \left(13(\cos(1.17601) + i\sin(1.17601))\right)^3$

       $= 2197(\cos((3)(1.17601)) + i\sin((3)(1.17601)))$

       $= 2197(\cos(3.52803) + i\sin(3.52803))$

       $= -2034.98809 - 828.02926i$

39. $z^3 = \left(25(\cos(2.21430) + i\sin(2.21430))\right)^3$

$\quad = 15,625(\cos((3)(2.21430)) + i\sin((3)(2.21430)))$

$\quad = 15,625(\cos(6.64290) + i\sin(6.64290))$

$\quad = 14,624.95769 + 5500.11251i$

41. $z^3 = \left(\sqrt{73}(\cos(-0.35877) + i\sin(-0.35877))\right)^3$

$\quad = \left(73^{3/2}\right)(\cos((3)(-0.35877)) + i\sin((3)(-0.35877)))$

$\quad = 623.71227(\cos(-1.07631) + i\sin(-1.07631))$

$\quad = 296.00110 - 548.99940i$

43. $z^3 = \left(\sqrt{17}(\cos(0.75597) + i\sin(0.75597))\right)^3$

$\quad = \left(17^{3/2}\right)(\cos((3)(0.75597)) + i\sin((3)(0.75597)))$

$\quad = 70.0928(\cos(2.26791) + i\sin(2.26791))$

$\quad = -45.00010 + 53.74004i$

45. $(1 + 2i)^0 = 1$ 　　　　　$(1 + 2i)^1 = 1 + 2i$ 　　　　　$(1 + 2i)^2 = -3 + 4i$

$(1 + 2i)^3 = -11 - 2i$ 　　　　　$(1 + 2i)^4 = -7 - 24i$

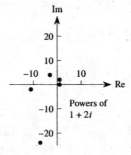

47. $z^4 = z^3 z$

$\quad = \left(\|z\|^3 (\cos 3\theta + i\sin 3\theta)\right)\left(\|z\|(\cos\theta + i\sin\theta)\right)$

$\quad = \|z\|^4 \left(\cos 3\theta \cos\theta + i^2 \sin 3\theta \sin\theta + i(\sin 3\theta \cos\theta + \cos 3\theta \sin\theta)\right)$

$\quad = \|z\|^4 \left([\cos 3\theta \cos\theta - \sin 3\theta \sin\theta] + i[\sin 3\theta \cos\theta + \cos 3\theta \sin\theta]\right)$

$\quad = \|z\|^4 (\cos(3\theta + \theta) + i\sin(3\theta + \theta))$

$\quad = \|z\|^4 (\cos 4\theta + i\sin 4\theta)$

49. **a.** $zw = (a + bi)(c + di) = (ac - bd) + (ad + bc)i$

b. If θ_1 and θ_2 are the angles associated with w and z, then

$zw = \|z\|(\cos\theta_1 + i\sin\theta_1)\|w\|(\cos\theta_2 + i\sin\theta_2)$

$\quad = \|z\|\|w\|(\cos\theta_1 \cos\theta_2 - \sin\theta_1 \sin\theta_2 + i(\sin\theta_1 \cos\theta_2 + \cos\theta_1 \sin\theta_2))$

$\quad = \|z\|\|w\|(\cos(\theta_1 + \theta_2) + i\sin(\theta_1 + \theta_2))$

c. This provides the extension of DeMoivre's theorem to products in trigonometric form:

$zw = \|z\|\|w\|(\cos(\theta_1 + \theta_2) + i\sin(\theta_1 + \theta_2))$

d. **(i)** $z = 1 + 2i$ and $w = 1 - 2i$. Since $\arctan(2) = 1.107$ radians, the trigonometric forms for z and w are

$$z = \sqrt{5}(\cos(1.107) + i\sin(1.107)) \text{ and } w = \sqrt{5}(\cos(-1.107) + i\sin(-1.107)). \text{ Then}$$

$$zw = \sqrt{5}\sqrt{5}(\cos(1.1 - 7 + (-1.107)) + i\sin(1.107 + (-1.107))$$

$$= 5(\cos(0) + i\sin(0)) = 5.$$

(ii) $z = \dfrac{1}{2} + \dfrac{\sqrt{3}}{2}i$ and $w = \dfrac{\sqrt{3}}{2} - \dfrac{1}{2}i$. Since $\arctan(\sqrt{3}) = \dfrac{\pi}{3}$ and $\arctan\left(\dfrac{1}{\sqrt{3}}\right) = \dfrac{\pi}{6}$, the trigonometric forms

for z and w are $z = 1\left(\cos\left(\dfrac{\pi}{3}\right) + i\sin\left(\dfrac{\pi}{3}\right)\right)$ and $w = 1\left(\cos\left(-\dfrac{\pi}{6}\right) + i\sin\left(-\dfrac{\pi}{6}\right)\right)$. Then

$$zw = (1 \cdot 1)\left(\cos\left(\dfrac{\pi}{3} + \left(-\dfrac{\pi}{6}\right)\right) + i\sin\left(\dfrac{\pi}{3} + \left(-\dfrac{\pi}{6}\right)\right)\right)$$

$$= \cos\left(\dfrac{\pi}{6}\right) + i\sin\left(\dfrac{\pi}{6}\right) = \dfrac{\sqrt{3}}{2} + \dfrac{1}{2}i.$$

51. **a.** If the angle associated with z is θ, then $\sqrt{z} = \sqrt{\|z\|}\left(\cos\left(\dfrac{\theta}{2}\right) + i\sin\left(\dfrac{\theta}{2}\right)\right)$. Like nonzero real numbers, nonzero complex numbers have two square roots, generated by the formula

$$\sqrt{z} = \sqrt{\|z\|}\left[\cos\left(\dfrac{\theta}{2}\right) + i\sin\left(\dfrac{\theta}{2}\right)\right] \text{ and } \sqrt{z} = \sqrt{\|z\|}\left[\cos\left(\dfrac{\theta}{2} + \pi\right) + i\sin\left(\dfrac{\theta}{2} + \pi\right)\right].$$

b. The trigonometric form for $z = \dfrac{1}{2} + \dfrac{\sqrt{3}}{2}i$ is $1\left(\cos\left(\dfrac{\pi}{3}\right) + i\sin\left(\dfrac{\pi}{3}\right)\right)$. Then

$$\sqrt{z} = \sqrt{1}\left(\cos\left(\dfrac{\pi}{6}\right) + i\sin\left(\dfrac{\pi}{6}\right)\right) = \dfrac{\sqrt{3}}{2} + \dfrac{1}{2}i \text{ and } \sqrt{z} = \sqrt{1}\left(\cos\left(\dfrac{7\pi}{6}\right) + i\sin\left(\dfrac{7\pi}{6}\right)\right) = -\dfrac{\sqrt{3}}{2} - \dfrac{1}{2}i.$$

c. $\left(\dfrac{\sqrt{3}}{2} + \dfrac{1}{2}i\right)^2 = \dfrac{3}{4} + 2\left(\dfrac{\sqrt{3}}{2}\right)\left(\dfrac{1}{2}\right)i - \dfrac{1}{4} = \dfrac{1}{2} + \dfrac{\sqrt{3}}{2}i$

d. The complete set of nth roots of z is given by

$$\sqrt[n]{\|z\|}\left(\cos\left(\dfrac{\theta + 2k\pi}{n}\right) + i\sin\left(\dfrac{\theta + 2k\pi}{n}\right)\right) \text{ for } k = 0, 1, 2, ..., n-1.$$

53. $z \cdot \overline{z} = (a + bi)(a - bi) = a^2 - (bi)^2 = a^2 + b^2 = \|a + bi\|^2 = \|z\|^2$

Section 8.4   The Road to Chaos

1. **a.** The limiting value, one of the equilibrium values $\dfrac{1 \pm \sqrt{1 - 4C}}{2}$, will be real whenever $C \leq 1/4$. Note however that for some $C \leq 1/4$, the equilibrium values may be the only starting values for which the iteration $x = f(x)$ where $f(x) = x^2 + C$ converges.

b. {0.5, 0.35, 0.2225, 0.14950625, 0.12235212, 0.11497004, 0.11321811, 0.11281834, 0.11272798, 0.1127076, 0.112703, ...}

c. The sequence of iterates diverges to infinity.

3. The equilibrium values for the difference equation are the solutions to the equation $x = f(x)$. The equilibrium values, $\dfrac{1 \pm \sqrt{1 - 4C}}{2}$, will be real whenever $C \leq 1/4$.

5. For every choice of C, the graph of $y = x$ intersects the shifted cubic $y = x^3 + C$.

7. The roots of $x = x + \cos x$ are the half-odd-integer multiples of π.

Chapter 8 Review Problems

1. Not an identity

2. Identity: $(\sin x \cos x)^2 + \sin^3 x = \sin x \left(\cos^2 x + \sin^2 x \right) = \sin x$

3. Not an identity

4. Not an identity

5. Identity: $(\sin x + \cos x)^2 = \sin^2 x + 2 \sin x \cos x + \cos^2 x = 1 + \sin 2x$

6. Identity: $\dfrac{\sin\theta}{1+\cos\theta} + \dfrac{\cos\theta}{\sin\theta} = \dfrac{\sin^2\theta + (1+\cos\theta)(\cos\theta)}{(1+\cos\theta)(\sin\theta)} = \dfrac{1+\cos\theta}{(1+\cos\theta)(\sin\theta)} = \dfrac{1}{\sin\theta}$

7. Identity: $\dfrac{1}{1+\cos t} + \dfrac{1}{1-\cos t} = \dfrac{1-\cos t + 1 + \cos t}{(1+\cos t)(1-\cos t)} = \dfrac{2}{1-\cos^2 t} = \dfrac{2}{\sin^2 t}$

8. Identity: $\cos^4\theta - \sin^4\theta - (\cos^2\theta + \sin^2\theta)(\cos^2\theta - \sin^2\theta) = \cos 2\theta$

9. Not an identity.

10. **a.** $\cos(x+y) + \cos(x-y) = \cos x \cos y - \sin x \sin y + \cos x \cos y + \sin x \sin y = 2\cos x \cos y$

    b. $\sin(x+y) + \sin(x-y) = \sin x \cos y + \cos x \sin y + \sin x \cos y - \cos x \sin y = 2\sin x \cos y$

11. With $T_3(x) = 3x - \dfrac{9}{2}x^3$, we have $T_3(0.2) = 0.564$. Below we plot $T_3(x)$ and $\sin 3x$.

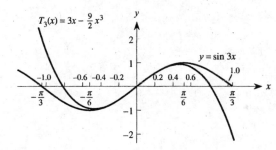

12. With $T_5(x) = 3x - \dfrac{9}{2}x^3 + \dfrac{81}{40}x^5$, we have $T_5(0.2) = 0.564648$. Below we plot $T_5(x)$ and $\sin 3x$.

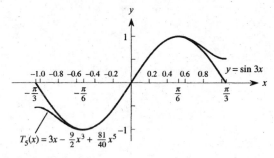

13. The maximum difference for $-1 \le x \le 1$ is about 0.06.

14. a. $z = -6 + 8i$; $\sqrt{(-6)^2 + 8^2} = 10$ and $\arctan\left(-\dfrac{8}{6}\right) = -0.927$. Since $-6 + 8i$ is in Quadrant II, the angle for z is

$-0.927 + \pi = 2.215$. Thus $z = 10(\cos 2.215 + i \sin 2.215)$.

$w = 5 - 2i$; $\sqrt{5^2 + (-2)^2} = \sqrt{29}$ and $\arctan\left(-\dfrac{2}{5}\right) = -0.381$. Since $5 - 2i$ is in Quadrant IV, the angle for

w is -0.381. Thus $w = \sqrt{29}(\cos(-0.381) + i \sin(-0.381))$.

b. Since $2.215 + (-0.381) = 1.834$,

$$\begin{aligned} zw &= (10)(\sqrt{29})(\cos 1.834 + i \sin 1.834) \\ &= 53.8516(\cos 1.834 + i \sin 1.834) \\ &= -14.01 + 51.997i \end{aligned}$$

Since $2.215 - (-0.381) = 2.596$,

$$\begin{aligned} \frac{z}{w} &= \frac{10}{\sqrt{29}}(\cos 2.596 + i \sin 2.596) \\ &= 1.85695(\cos 2.596 + i \sin 2.596) \\ &= -1.587 + 0.964i \end{aligned}$$

For $\dfrac{w}{z}$ we take the reciprocal of the modulus and the negative of the angle, getting:

$$\begin{aligned} \frac{w}{z} &= \frac{\sqrt{29}}{10}(\cos(-2.596) + i \sin(-2.596)) \\ &= 0.53852(\cos(-2.596) + i \sin(-2.596)) \\ &= -0.46 - 0.279i \end{aligned}$$

15. a. $z = 3(\cos 52° + i \sin 52°)$

b. $z = 1.847 + 2.364i$

c. $z^5 = 3^5(\cos 260° + i \sin 260°) = -42.197 - 239.308i$

d. $\sqrt{z} = \sqrt{3}(\cos 26° + i \sin 26°) = 1.557 + 0.759i$

16.

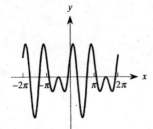

Period 2π

17.

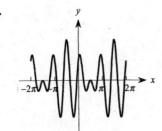

Period 2π

18.

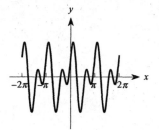

Period π

19.

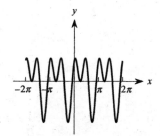

Period π

20.

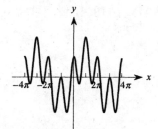

Period 4π

21.

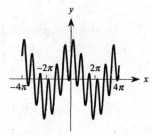

Period 4π

22. For integers m and n, the frequency of $\sin mx + \cos nx$ is $\dfrac{m \cdot n}{\text{LCM}(m,n)}$, where LCM is the least common

multiple. Thus the period is $\dfrac{2\pi \, \text{LCM}(m,n)}{m \cdot n}$. If the two coefficients of x are of the form m and $\dfrac{1}{n}$ for

integers m and n, the period is $2\pi n$.

Chapter 9   *Geometric Models*

Section 9.2   Analytic Geometry

1. The distance between $P(2, 4)$ and $Q(5, 8)$ is $\sqrt{(5-2)^2+(8-4)^2}=5$.

3. The distance between $P(-4, 1)$ and $Q(0, 4)$ is $\sqrt{(0-(-4))^2+(4-1)^2}=5$.

5. The distance between $P(-1, 5)$ and $Q(3, 7)$ is $\sqrt{(3-(-1))^2+(7-4)^2}=\sqrt{20}\approx 4.47$.

7. The midpoint of the segment connecting $P(2, 4)$ and $Q(5, 8)$ is $\left(\frac{5+2}{2},\frac{8+4}{2}\right)=(3.5, 6)$.

9. The point $\frac{1}{3}$ of the way from $P(-4, 1)$ to $Q(0, 4)$ is $\left(-4+\frac{1}{3}[0-(-4)], 1+\frac{1}{3}[4-1]\right)=\left(-\frac{8}{3}, 2\right)$.

11. The radius of the given circle is $r=\sqrt{(5-8)^2+(2-(-2))^2}=5$. Thus, an equation of the given circle is $(x-5)^2+(y-2)^2=25$.

13. The center of the given circle is the midpoint of the diameter, which is $\left(\frac{2+10}{2},\frac{4+4}{2}\right)=(6, 4)$. The radius of the circle is $\frac{1}{2}\sqrt{(10-2)^2+(4-4)^2}=4$. Thus, an equation of the given circle is $(x-6)^2+(y-4)^2=16$.

15. If (x, y) is a point on the circle with $P(2, 4)$ and $Q(10, 4)$ as endpoints of a diameter, the slope of the line that joins (x, y) to $P(2, 4)$ is $(y-4)/(x-2)$, and the slope of the line joint (x, y) to $Q(10, 4)$ is $(y-4)/(x-10)$. Since these two lines are perpendicular, the product of their slopes is -1. So,

$$\frac{y-4}{x-2}\cdot\frac{y-4}{x-10}=-1.$$

Simplifying and completing the square in x, we get:

$$(y-4)^2=-(x-2)(x-10)$$
$$x^2-12x+20+(y-4)^2=0$$
$$x^2-12x+20+16+(y-4)^2=0+16$$
$$(x-6)^2+(y-4)^2=16$$

Similarly, any point (x, y) on the circle with diameter having endpoints $P(2, 4)$ and $Q(10, 14)$ satisfies the equation

$$\frac{y-4}{x-2}\cdot\frac{y-14}{x-10}=-1.$$

Simplifying and completing the square in x and y, we get:

$$(y-4)(y-14)=-(x-2)(x-10)$$
$$x^2-12x+20+y^2-18y+56=0$$
$$x^2-12x+20+16+y^2-18y+56+25=0+16+25$$
$$x^2-12x+36+y^2-18y+81=41$$
$$(x-6)^2+(y-9)^2=41$$

17. By completing the square in x and y, we get:

$$x^2 + 4x + y^2 + 6y = 12$$
$$x^2 + 4x + 4 + y^2 + 6y + 9 = 12 + 4 + 9$$
$$(x+2)^2 + (y+3)^2 = 25$$

This equation represents a circle with center at $(-2, -3)$ and radius 5.

19. By completing the square in x and y, we get:

$$x^2 + 10x + y^2 - 4y = 71$$
$$x^2 + 10x + 25 + y^2 - 4y + 4 = 71 + 25 + 4$$
$$(x+5)^2 + (y-2)^2 = 100$$

This equation represents a circle with center at $(-5, 2)$ and radius 10.

21. By completing the square in x and y, we get:

$$x^2 - 2x + y^2 + 6y + 6 = 0$$
$$x^2 - 2x + y^2 + 6y = -6$$
$$x^2 - 2x + 1 + y^2 + 6y + 9 = -6 + 1 + 9$$
$$(x-1)^2 + (y+3)^2 = 4$$

This equation represents a circle with center at $(1, -3)$ and radius 2.

23. **a.**

t	-2	-1	0	1	2	3	4	5
x	-1	1	3	5	7	9	11	13
y	14	9	4	-1	-6	-11	-16	-21

b.

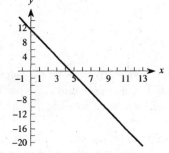

c. The line has slope $-\frac{5}{2}$

d. An equation of the line is $y - 14 = -\frac{5}{2}(x+1) \implies y = -\frac{5}{2}x - \frac{5}{2} + 14 \implies y = -\frac{5}{2}x + \frac{23}{2}$

e. The midpoints are $(0, 11.5)$, $(2, 6.5)$, $(4, 1.5)$, $(6, -3.5)$, $(8, -8.5)$, $(10, -13.5)$, and $(12, -18.5)$. The following table gives the values using the parametric equations.

t	-1.5	-0.5	0.5	1.5	2.5	3.5	4.5
x	0	2	4	6	8	10	12
y	11.5	6.5	1.5	-3.5	-8.5	-13.5	-18.5

The results are the same because the values of t are the midpoints of the given intervals of t.

25. $x = x_0 + t(x_1 - x_0)$, so $t = (x - x_0)/(x_1 - x_0)$

$y = y_0 + t(y_1 - y_0)$, so $t = (y - y_0)/(y_1 - y_0)$

Therefore,

$$\frac{y - y_0}{y_1 - y_0} = \frac{x - x_0}{x_1 - x_0}$$

$$y - y_0 = \frac{y_1 - y_0}{x_1 - x_0}(x - x_0)$$

$$y - y_0 = m(x - x_0)$$

where $\dfrac{y_1 - y_0}{x_1 - x_0} = m = $ slope of the line.

27. Substituting $x = 1 + 2t$ and $y = 2 - t$ into $x^2 + y^2 = 25$, we get, after simplification, the equation $5t^2 - 20 = 0$. Substituting the solutions to this equation ($t = 2$ and $t = -2$) into the parametric form of the line, we get $(5, 0)$ and $(-3, 4)$ as the two points of intersection of the line and the circle.

Exercising Your Algebra Skills

1. $x^2 + 8x + 25 = (x^2 + 8x + 16) + (25 - 16) = (x + 4)^2 + 9$

2. $x^2 - 8x + 25 = (x^2 - 8x + 16) + (25 - 16) = (x - 4)^2 + 9$

3. $x^2 - 6x + 5 = (x^2 - 6x + 9) + (5 - 9) = (x - 3)^2 - 4$

4. $x^2 + 6x + 5 = (x^2 + 6x + 9) + (5 - 9) = (x + 3)^2 - 4$

5. $y^2 + 10y + 26 = (y^2 + 10y + 25) + (26 - 25) = (y + 5)^2 + 1$

6. $y^2 - 10y + 26 = (y^2 - 10y + 25) + (26 - 25) = (y - 5)^2 + 1$

7. $y^2 + 4y - 12 = (y^2 + 4y + 4) + (-12 - 4) = (y + 2)^2 - 16$

8. $y^2 - 4y - 12 = (y^2 - 4y + 4) + (-12 - 4) = (y - 2)^2 - 16$

Section 9.3   Conic Sections: The Ellipse

1. **a.** Since the gravitational force is proportional to $1/r^2$, the greater the distance, the smaller the force. Thus the gravitational force will be largest when the satellite is closest to Earth at perihelion (the point on the right) and the force will be smallest at aphelion (the point on the left) when the satellite is at its farthest point from Earth.

   b. The satellite will accelerate at an increasingly greater rate as the satellite gets nearer Earth reaching its maximum speed at perihelion (the point on the right) and then decelerates as it moves away from Earth.

   c. The satellite will lose velocity until it reaches aphelion (the point on the left) where the speed is least and then begin accelerating again.

3. The satellite will no longer stay in orbit and will be drawn steadily closer to Earth by the pull of gravity in a spiral orbit, as pictured, eventually either burning up in the atmosphere or crashing into Earth.

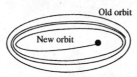

5. The most circular orbit corresponds to the one in which the perihelion and the aphelion are the closest, since in that case c is the closest to 0. Venus' orbit is the most circular and Pluto's the least circular.

7. From $2a = 69.8 + 46 = 115.8$, we get $a = 57.9$. Further, we have $a = c + 46$, which gives $c = 11.9$. Finally, from $a^2 = b^2 + c^2$ we get $b = 56.66$. The equation of the orbit of Mercury is:

$$\frac{x^2}{(57.9)^2} + \frac{y^2}{(56.66)^2} = 1, \text{ or } \frac{x^2}{3352.41} + \frac{y^2}{3210.36} = 1$$

9. **a.** With $a = 620/2 = 310$ and $b = 513/2 = 256.5$, and taking the x-axis to be oriented along the major axis, we have the equation

    $$\frac{x^2}{96,100} + \frac{y^2}{65,792.25} = 1.$$

    b. Each focus is a distance c from the center along the major axis, so the distance between foci is $2c$. Since $c^2 = a^2 - b^2$, we have $2c = 2\sqrt{96,100 - 65,792.25} = 348.182$ ft.

    c. The center of an ellipse and any two adjacent vertices determine a right triangle with legs of length a and b. The distance between adjacent vertices is then $\sqrt{a^2 + b^2} = \sqrt{96,100 + 65,792.25} = 402.358$ ft.

11. In the standard form the equation is:

    $$\frac{(x-0)^2}{\left(\frac{1}{2}\right)^2} + \frac{(y-0)^2}{\left(\frac{1}{3}\right)^2} = 1,$$

    so $a = \frac{1}{2}$, $b = \frac{1}{3}$, and $c = \sqrt{\left(\frac{1}{2}\right)^2 + \left(\frac{1}{3}\right)^2} = \frac{\sqrt{5}}{6}$. The center is at $(0, 0)$, and the major axis is along the x-axis. The vertices are $\left(-\frac{1}{2}, 0\right), \left(\frac{1}{2}, 0\right), \left(0, -\frac{1}{3}\right), \left(0, \frac{1}{3}\right)$, and the foci are at $\left(-\frac{\sqrt{5}}{6}, 0\right)$ and $\left(\frac{\sqrt{5}}{6}, 0\right)$.

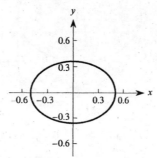

13.
$$x^2 + 4y^2 + 2x + 8y = -1$$
$$(x^2 + 2x) + 4(y^2 + 2y) = -1$$
$$(x^2 + 2x + 1) + 4(y^2 + 2y + 1) = -1 + 1 + 4 \cdot 1$$
$$(x+1)^2 + 4(y+1)^2 = 4$$
$$\frac{(x+1)^2}{4} + \frac{(y+1)^2}{1} = 1$$

This is the equation of an ellipse with center at $(-1, -1)$, major axis parallel to the x-axis, $a = 2$, $b = 1$, and $c = \sqrt{a^2 - b^2} = \sqrt{2^2 - 1^2} = \sqrt{3}$. The vertices are $(-3, -1)$, $(1, -1)$, $(-1, -2)$, $(-1, 0)$, and the foci are at $(-1 - \sqrt{3}, -1)$ and $(-1 + \sqrt{3}, -1)$.

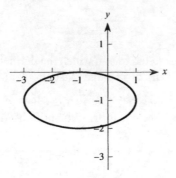

15.
$$x^2 + 4y^2 + 20x - 40y + 100 = 0$$
$$(x^2 + 20x) + 4(y^2 - 10y) = -100$$
$$(x^2 + 20x + 100) + 4(y^2 - 10y + 25) = -100 + 100 + 4 \cdot 25$$
$$(x + 10)^2 + 4(y - 5)^2 = 100$$
$$\frac{(x + 10)^2}{100} + \frac{(y - 5)^2}{25} = 1$$

This is the equation of an ellipse with center at $(-10, 5)$, major axis parallel to the x-axis, $a = 10$, $b = 5$, and $c = \sqrt{a^2 - b^2} = \sqrt{10^2 - 5^2} = 5\sqrt{3}$. The vertices are $(-20, 5)$, $(0, 5)$, $(-10, 0)$, $(-10, 10)$, and the foci are at $(-10 - 5\sqrt{3}, 5)$ and $(-10 + 5\sqrt{3}, 5)$. The graph is shown below.

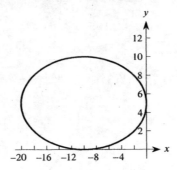

17.
$$9x^2 + y^2 - 54x + 4y = -76$$
$$9(x^2 - 6x) + (y^2 + 4y) = -76$$
$$9(x^2 - 6x + 9) + (y^2 + 4y + 4) = -76 + 9 \cdot 9 + 4$$
$$9(x - 3)^2 + (y + 2)^2 = 9$$
$$\frac{(x - 3)^2}{1} + \frac{(y + 2)^2}{9} = 1$$

This is the equation of an ellipse with center at $(3, -2)$, major axis parallel to the y-axis, $a = 1$, $b = 3$, and $c = \sqrt{b^2 - a^2} = \sqrt{3^2 - 1^2} = 2\sqrt{2}$. The vertices are $(3, -5)$, $(3, 1)$, $(2, -2)$, $(4, -2)$, and the foci are at $(3, -2 - 2\sqrt{2})$ and $(3, -2 + 2\sqrt{2})$.

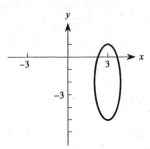

19.

$$\sqrt{(x-c)^2+y^2} = 2a - \sqrt{(x+c)^2+y^2}$$ Square both sides.

$$(x-c)^2+y^2 = 4a^2 - 4a\sqrt{(x+c)^2+y^2} + (x+c)^2+y^2$$ Isolate radical.

$$4a\sqrt{(x+c)^2+y^2} = 4a^2 + (x+c)^2 - (x-c)^2$$ Collect like terms.

$$4a\sqrt{(x+c)^2+y^2} = 4a^2 + 4cx$$ Divide both sides by 4.

$$a\sqrt{(x+c)^2+y^2} = a^2 + cx$$ Square both sides.

$$a^2(x^2 + 2xc + c^2 + y^2) = a^4 + 2a^2cx + c^2x^2$$ Collect like terms with variables on the left.

$$a^2x^2 - c^2x^2 + a^2y^2 = a^4 - a^2c^2$$ Substitute $b^2 = a^2 - c^2$.

$$b^2x^2 + a^2y^2 = a^2b^2$$ Divide both sides by a^2b^2.

$$\frac{x^2}{a^2} + \frac{y^2}{b^2} = 1$$

Section 9.4   Conic Sections: The Hyperbola and the Parabola

1. **a.** The equation of a parabola with vertex at the origin, axis of symmetry oriented along the positive y-axis and focus at distance 6 from the vertex is $y = x^2/(4 \cdot 6)$, or $y = x^2/24$.

   b. In terms of our equation, the diameter associated with a point (x, y) on the parabola is twice the x-coordinate. If, as we are given, the y-coordinate is known to be 6, then solving $6 = x^2/24$ gives $x = 12$. The diameter of the extended dish would be 24 inches.

3. **a.** Rewriting the equation of the hyperbola as $\dfrac{x^2}{a^2} = 1 + \dfrac{y^2}{b^2}$, it is easy to see that when x and y are

extremely large, $\dfrac{y^2}{b^2}$ is extremely large and that adding 1 is essentially meaningless. So, the equation

reduces to $\dfrac{x^2}{a^2} = \dfrac{y^2}{b^2}$.

   b. Substituting 0 for 1 on the right of the given equation and factoring the left-hand side as a difference of

squares, we have $\left(\dfrac{x}{a} - \dfrac{y}{b}\right)\left(\dfrac{x}{a} + \dfrac{y}{b}\right) = 0$. This gives the two lines $y = \dfrac{b}{a}x$ and $y = -\dfrac{b}{a}x$.

   c. The slopes are $\dfrac{b}{a}$ and $-\dfrac{b}{a}$.

5.
$$x^2 + 4y^2 + 2x + 8y = 11$$
$$(x^2 + 2x) + 4(y^2 + 2y) = 11$$
$$(x^2 + 2x + 1) + 4(y^2 + 2y + 1) = 11 + 1 + 4 \cdot 1$$
$$(x+1)^2 + 4(y+1)^2 = 16$$
$$\frac{(x+1)^2}{16} + \frac{(y+1)^2}{4} = 1$$

This is the equation of an ellipse with center at $(-1, -1)$, major axis parallel to the x-axis, $a = 4$, $b = 2$, and $c = \sqrt{a^2 - b^2} = \sqrt{4^2 - 2^2} = \sqrt{12}$. The vertices are $(-5, -1)$, $(3, -1)$, $(-1, -3)$, $(-1, 1)$, and the foci are at $(-1 - \sqrt{12}, -1)$ and $(-1 + \sqrt{12}, -1)$.

7.
$$x^2 - 4y^2 + 2x - 8y = 19$$
$$(x^2 + 2x) - 4(y^2 + 2y) = 19$$
$$(x^2 + 2x + 1) - 4(y^2 + 2y + 1) = 19 + 1 - 4 \cdot 1$$
$$(x+1)^2 - 4(y+1)^2 = 16$$
$$\frac{(x+1)^2}{16} - \frac{(y+1)^2}{4} = 1$$

This is the equation of a hyperbola with center at $(-1, -1)$, axis parallel to the x-axis, $a = 4$, $b = 2$, and $c = \sqrt{a^2 + b^2} = \sqrt{4^2 + 2^2} = \sqrt{20}$. The vertices are $(-5, -1)$ and $(3, -1)$, and the foci are at $(-1 - \sqrt{20}, -1)$ and $(-1 + \sqrt{20}, -1)$.

9.
$$4x^2 - 9y^2 - 16x - 18y = 31$$
$$4(x^2 - 4x) - 9(y^2 + 2y) = 31$$
$$4(x^2 - 4x + 4) - 9(y^2 + 2y + 1) = 31 + 4 \cdot 4 - 9 \cdot 1$$
$$4(x-2)^2 - 9(y+1)^2 = 38$$
$$\frac{(x-2)^2}{38/4} - \frac{(y+1)^2}{38/9} = 1$$

This is the equation of a hyperbola with center at $(2, -1)$, axis parallel to the x-axis, $a = \sqrt{38}/2$, $b = \sqrt{38}/3$, and $c = \sqrt{a^2 + b^2} = \sqrt{(\sqrt{38}/2)^2 + (\sqrt{38}/3)^2} = \sqrt{494}/6$. The vertices are $(2 - \sqrt{38}/2, -1)$ and $(2 + \sqrt{38}/2, -1)$, and the foci are at $(2 - \sqrt{494}/6, -1)$ and $(2 + \sqrt{494}/6, -1)$.

11.
$$4x^2 + 4y^2 - 24x + 16y + 43 = 0$$
$$4(x^2 - 6x) + 4(y^2 + 4y) = -43$$
$$4(x^2 - 6x + 9) + 4(y^2 + 4y + 4) = -43 + 4 \cdot 9 + 4 \cdot 4$$
$$4(x-3)^2 + 4(y+2)^2 = 9$$
$$(x-3)^2 + (y+2)^2 = \frac{9}{4}$$

This is the equation of a circle with center at $(3, -2)$ and radius $\frac{3}{2}$.

13.
$$9x^2 - 4y^2 + 18x - 16y = 6$$
$$9(x^2 + 2x) - 4(y^2 + 4y) = 6$$
$$9(x^2 + 2x + 1) - 4(y^2 + 4y + 4) = 6 + 9 \cdot 1 - 4 \cdot 4$$
$$9(x+1)^2 - 4(y+2)^2 = -1$$
$$\frac{(y+2)^2}{1/4} - \frac{(x+1)^2}{1/9} = 1$$

This is the equation of a hyperbola with center at $(-1, -2)$, axis parallel to the y-axis, $a = 1/2$, $b = 1/3$, and $c = \sqrt{a^2 + b^2} = \sqrt{(1/3)^2 + (1/2)^2} = \sqrt{13}/6$. The vertices are $(-1, -\frac{5}{2})$ and $(-1, -\frac{3}{2})$, and the foci are at $(-1, -\frac{7}{3})$ and $(-1, -\frac{5}{3})$.

15. **a.** The slope of the displayed lines are -4, $-9/4$, -1, $-4/9$, and $-1/4$.
    b. Slope is given by $m = -1/x^2$.

Section 9.5   Parametric Curves

1. **a.**

t	-2	-1	0	1	2
$x = 4 - 3t$	10	7	4	1	-2
$y = 2 - 5t$	12	7	2	-3	-8

Using any two points from the table, say $(4, 2)$ and $(1, -3)$, the slope is computed to be $(-3 - 2)/(1 - 4) = -5/-3 = 5/3$.

b. Using the slope $5/3$ from part (a) and the point $(1, -3)$, an equation for the line is
$$y - (-3) = \tfrac{5}{3}(x - 1), \quad \text{or } y = \tfrac{5}{3}x - \tfrac{14}{3}.$$

c. The slope of the line, $5/3$, is equal to the ratio of the coefficient of t in the expression for y, which is -5, to the corresponding coefficient in the expression for x, which is -3.

d. Solving for t in $x = 4 - 3t$ gives $t = -\tfrac{1}{3}x + \tfrac{4}{3}$. Substituting into $y = 2 - 5t$ gives
$$y = 2 - 5\left(-\tfrac{1}{3}x + \tfrac{4}{3}\right) \Rightarrow y = \tfrac{5}{3}x - \tfrac{14}{3}.$$

e. Solving of t in $y = 2 - 5t$ gives $t = -\tfrac{1}{5}y + \tfrac{2}{5}$. Substituting into $x = 4 - 3t$ gives
$$x = 4 - 3\left(-\tfrac{1}{5}y + \tfrac{2}{5}\right) \Rightarrow x = \tfrac{3}{5}y + \tfrac{14}{5}.$$

Note that $x = \tfrac{3}{5}y + \tfrac{14}{5} \Rightarrow y = \tfrac{5}{3}x - \tfrac{14}{3}.$

3. **a.**

t	-2	-1.5	-1	-0.5	0	0.5	1	1.5	2
$x = t^3 + 1$	-7	-2.375	0	0.875	1	1.125	2	4.375	9
$y = t^2 - 2$	2	0.25	-1	-1.75	-2	-1.75	-1	0.25	2

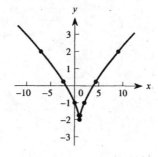

b. A function grapher using PARAMETRIC mode might produce a graph like the graph shown below. This graph is a smoother version of the graph in part (a).

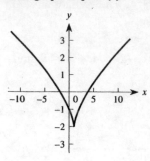

c. Solving for t in $x = t^3 + 1$ gives $t = (x-1)^{1/3}$. Substituting into $y = t^2 - 2$ gives $y = (x-1)^{2/3} - 2$.

d. Graphs created on a function grapher using the FUNCTION mode will vary according to the function grapher's ability to handle cube roots. If the function grapher does not compute or gives complex cube roots of negative numbers, then the grapher may draw the curve only for $x > 1$, as shown below.

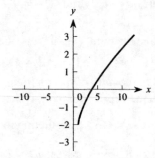

5. For $a = 1$, $b = 3$; period $= 6\pi$

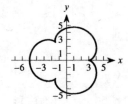

For $a = 1$, $b = 4$; period $= 8\pi$

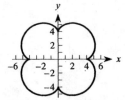

For $a = 1$, $b = 5$; period $= 10\pi$

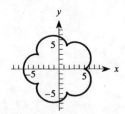

For $a = 1$, $b = 6$; period $= 12\pi$

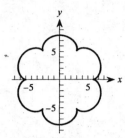

The number of loops appears to be b, and the period seems to be $2b\pi$.

7. For $a = 2$, $b = 3$; period $= 6\pi$

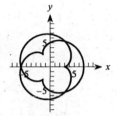

For $a = 2$, $b = 5$; period $= 10\pi$

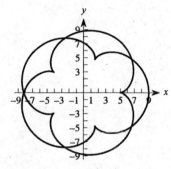

For $a = 1$, $b = 7$; period $= 14\pi$

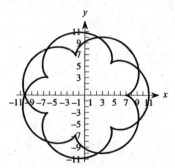

For $a = 1$, $b = 9$; period $= 18\pi$

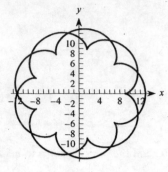

The number of loops appears to be b and the period seems to be $2b\pi$.

9. a. For $a = 1$ and $b = 2$: $x = 2\cos t$, $y = 0$

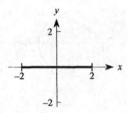

b. For $a = 1$ and $b = 3$: $x = 2\cos t + \cos 2t$, $y = 2\sin t - \sin 2t$

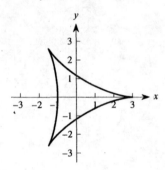

c. For $a = 1$, $b = 4$: $x = 3\cos t + \cos 3t$, $y = 3\sin t - \sin 3t$

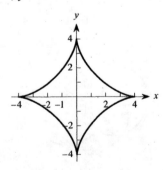

d. For $a = 2, b = 3$: $x = \cos t + 2\cos\frac{1}{2}t$, $y = \sin t - 2\sin\frac{1}{2}t$

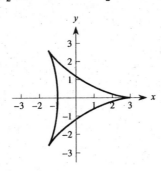

11. Solving the equation $y = a - a\cos t$ for t gives

$$t = \arccos\left(\frac{a-y}{a}\right)$$

Substituting for t into $x = at - a\sin t$ gives

$$x = a\arccos\left(\frac{a-y}{a}\right) - a\sin\left(\arccos\left(\frac{a-y}{a}\right)\right).$$

Section 9.6   The Polar Coordinate System

1. **a.** $P\left(5, \frac{\pi}{6}\right)$ $Q\left(3, \frac{2\pi}{3}\right)$ $R\left(6, \frac{7\pi}{6}\right)$ $S\left(2, \frac{5\pi}{3}\right)$

   b. $P\left(5, -\frac{11\pi}{6}\right)$ $Q\left(3, -\frac{4\pi}{3}\right)$ $R\left(6, -\frac{5\pi}{6}\right)$ $S\left(2, -\frac{\pi}{3}\right)$

   c. $P\left(-5, \frac{7\pi}{6}\right)$ $Q\left(-3, \frac{5\pi}{3}\right)$ $R\left(-6, \frac{\pi}{6}\right)$ $S\left(-2, \frac{2\pi}{3}\right)$

   d. $P\left(-5, -\frac{5\pi}{6}\right)$ $Q\left(-3, -\frac{\pi}{3}\right)$ $R\left(-6, -\frac{11\pi}{6}\right)$ $S\left(-2, -\frac{4\pi}{3}\right)$

3. **a.** Converting $P(4, 4)$ to polar coordinates we find that $r^2 = (4)^2 + (4)^2 = 32$, and that θ lies in the first quadrant with $\tan\theta = 4/4 = 1$. Thus $r = \sqrt{32}$ and $\theta = \pi/4$.

   b. Converting $P(-4, 4)$ to polar coordinates we find that $r^2 = (-4)^2 + (4)^2 = 32$, and that θ lies in the second quadrant with $\tan\theta = 4/-4 = -1$. Thus $r = \sqrt{32}$ and $\theta = -\pi/4 + \pi = 3\pi/4$.

   c. Converting $P(-4, -4)$ to polar coordinates we find that $r^2 = (-4)^2 + (-4)^2 = 32$, and that θ lies in the third quadrant with $\tan\theta = -4/-4 = 1$. Thus $r = \sqrt{32}$ and $\theta = \pi/4 + \pi = 5\pi/4$.

   d. Converting $P(4, -4)$ to polar coordinates we find that $r^2 = (4)^2 + (-4)^2 = 32$, and that θ lies in the fourth quadrant with $\tan\theta = -4/4 = -1$. Thus $r = \sqrt{32}$ and $\theta = -\pi/4$.

   e. Converting $P(3, -4)$ to polar coordinates we find that $r^2 = (3)^2 + (-4)^2 = 25$, and that θ lies in the fourth quadrant with $\tan\theta = -4/3$. Thus $r = 5$ and $\theta = \arctan(-4/3)$.

   f. Converting $P(-3, 4)$ to polar coordinates we find that $r^2 = (-3)^2 + (4)^2 = 25$, and that θ lies in the second quadrant with $\tan\theta = 4/-3$. Thus $r = 5$ and $\theta = \arctan(-4/3) + \pi$.

   g. Converting $P(8, 3)$ to polar coordinates we find that $r^2 = (8)^2 + (3)^2 = 73$, and that θ lies in the first quadrant with $\tan\theta = 3/8$. Thus $r = \sqrt{73}$ and $\theta = \arctan(3/8)$.

   h. Converting $P(3, 8)$ to polar coordinates we find that $r^2 = (3)^2 + (8)^2 = 73$, and that θ lies in the first quadrant with $\tan\theta = 8/3$. Thus $r = \sqrt{73}$ and $\theta = \arctan(8/3)$.

5. **a.** $r = 4000$ and $\theta = (3000/24{,}000)2\pi = \pi/4$. That is, $(4000, \pi/4)$.

   b. $r = 4000$ and $\theta = (7500/24{,}000)2\pi = 5\pi/8$. That is, $(4000, 5\pi/8)$.

   c. $r = 4000 + 22{,}800 = 26{,}800$ and $\theta = (-5200/24{,}000)2\pi = -13\pi/30$. That is, $(26{,}800, -13\pi/30)$.

Section 9.7   Families of Curves in Polar Coordinates

1.

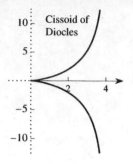

3.

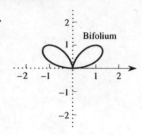

3.

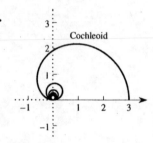

5.

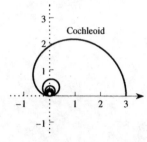

5.

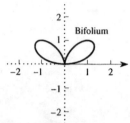

7.

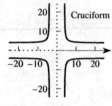

7.

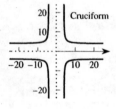

9.

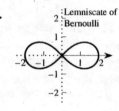

11.

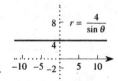

(Note: When $\theta = 0$, r is undefined.)

Problems 13–27 call for experimentation. Responses will vary significantly. Some possible answers are given.

13.   Changing the numerator c in $c/\sin \theta$ to a different value influences the loop near the origin. When $|c| > 2$, there is no loop. For $|c| < 2$, , the loop increases in size as c nears 0, and the two pieces of the curve approach each other, nearing a circle when $c = 0$. Replacing sin with cos results in the curve being rotated $90°$ counterclockwise.

15.   The coefficient whose value is 5 controls the scale of the plot, most visibly the extent of the loop. Changing the coefficient of cos θ has a similar effect. Increasing the numerator in the $(1/\cos \theta)$ term causes the loop to shrink to nothing and the position of the curve's cusp to move left on the x-axis. Decreasing the numerator causes the loop and the whiskers to retract until the numerator becomes 0 at which point the curve is a circle. Replacing cos with sin results in the curve being rotated 90° counterclockwise.

17. The constant a controls the radial scale of the graph, not its shape. Replacing a with its negative will rotate the curve 180°.

19. The numerator controls the radial scale of the graph. The constants in the denominator control how circular the ellipse is. If the coefficient of cos θ is large enough, the curve will become a hyperbola. Replacing cos with sin results in the graph being rotated 90° counterclockwise.

21. The coefficient of cos 3θ controls the radial scale of the graph. The coefficient n of θ determines the number of petals on the graph. If n is an odd integer, there will be n petals. If n is an even integer, there will be $2n$ petals. If n is not an integer, petals may overlap. Replacing cos with sin results in the graph being rotated 30° counterclockwise.

23. a. Assuming that a/b already is in lowest terms (that is, a and b have no common factors), then the number of loops will be a if a is odd and will be $2a$ if a is even.

    b. The range from $\frac{b}{a}\left(\frac{\pi}{2}\right)$ to $\frac{b}{a}\left(\frac{3\pi}{2}\right)$ traces one loop.

    c. If a is even, then the range from 0 to $2\pi b$ is needed to draw the entire curve. If a is odd, then the range from 0 to πb suffices.

25. There will be n loops if n is an even integer and $2n$ loops is n is odd. The loops are evenly spaced around a circle centered along the angles $\left(\frac{2\pi}{n}\right)k$ for $k = 0, 1, 2, \ldots, n - 1$ if n is even and $\left(\frac{\pi}{n}\right)k$ for $k = 0, 1, 2, \ldots, 2n - 1$ if n is odd.

27. If $a = b$, there will be n loops spaced equally within a circle. If $a < b$, there are n loops alternating with n shorter loops all arranged equally spaced within a circle. If $a > b$, there are no loops, only bumps arranged on a circle. There will be n such bumps.

Chapter 9 Review Problems

1. a. The center of the ellipse is $(1, -2)$ with $a = 3$ and $b = 5$. An equation for the ellipse is:
$$\frac{(x-1)^2}{9} + \frac{(y+2)^2}{25} = 1$$

   b. The center of the ellipse is $(1, -2)$ with $a = 5$ and $b = 3$. An equation for the ellipse is:
$$\frac{(x-1)^2}{25} + \frac{(y+2)^2}{9} = 1$$

2. a. The center of the hyperbola is $(2, 0)$ with axis parallel to the y-axis; $a = 3$ and $c = 5$, so $b = 4$. An equation for the hyperbola is:
$$\frac{y^2}{9} - \frac{(x-2)^2}{16} = 1$$

   b. The center of the hyperbola is $(0, -1)$ with axis parallel to the x-axis; $a = 2$ and $c = \sqrt{13}$, so $b = 3$. An equation for the hyperbola is:
$$\frac{x^2}{4} - \frac{(y+1)^2}{9} = 1$$

3. a. The vertex of the parabola is $(-4, 3)$ and $c = 1$. Since the parabola opens vertically, an equation for the parabola is:
$$y - 3 = \frac{(x+4)^2}{4(1)}, \text{ or } y = \frac{(x+4)^2}{4} + 3$$

b. The vertex of the parabola is $(-1, 3)$ and $c = 2$. Since the parabola opens horizontally, an equation for the parabola is:

$$x+1 = \frac{(y-3)^2}{4(2)}, \quad \text{or} \quad x = \frac{(y-3)^2}{8} - 1$$

4. The graph of $xy = 5$ is a hyperbola with axis on the line $y = x$.

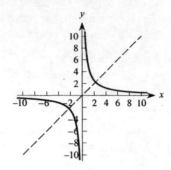

5. $x^2 - 6x + y - 34 = 0$

$$x^2 - 6x = -y + 34$$
$$x^2 - 6x + 9 = -y + 34 + 9$$
$$(x-3)^2 = -y + 43$$
$$y - 43 = -(x-3)^2$$

The conic is a parabola opening downward with vertex $(3, 43)$. Since $4c = -1 \Rightarrow$ $c = -0.25$, the focus is at $(3, 42.75)$.

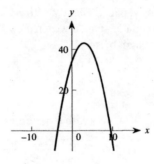

6. $x^2 + y^2 - 8x + 6y + 9 = 0$

$$(x^2 - 8x) + (y^2 + 6y) = -9$$
$$(x^2 - 8x + 16) + (y^2 + 6y + 9) = -9 + 16 + 9$$
$$(x-4)^2 + (y+3)^2 = 16$$

The conic is a circle with center $(4, -3)$ and radius 4.

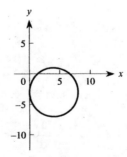

7.
$$2x^2 + 3y^2 + 20x - 12y + 28 = 0$$
$$2(2x^2 + 10x) + 3(y^2 - 4y) = -28$$
$$2(x^2 + 10x + 25) + 3(y^2 - 4y + 4) = -28 + 2 \cdot 25 + 3 \cdot 4$$
$$2(x+5)^2 + 3(y-2)^2 = 34$$
$$\frac{(x+5)^2}{17} + \frac{(y-2)^2}{34/3} = 1$$

The conic is an ellipse with center at $(-5, 2)$, major axis parallel to the x-axis, and $a = \sqrt{17}$ and $b = \sqrt{34/3}$.

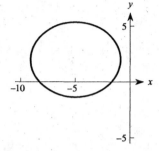

8.
$$3x^2 - 4y^2 - 6x - 24y - 45 = 0$$
$$3(x^2 - 2x) - 4(y^2 + 6y) = 45$$
$$3(x^2 - 2x + 1) - 4(y^2 + 6y + 9) = 45 + 3 \cdot 1 - 4 \cdot 9$$
$$3(x-1)^2 - 4(y+3)^2 = 12$$
$$\frac{(x-1)^2}{4} - \frac{(y+3)^2}{3} = 1$$

The conic is a hyperbola with center at $(1, -3)$, axis parallel to the x-axis, and $a = 2$ and $b = \sqrt{3}$.

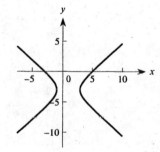

9.
$$3y^2 - 2x^2 - 12y + 12x - 24 = 0$$
$$3(y^2 - 4y) - 2(x^2 - 6x) = 24$$
$$3(y^2 - 4y + 4) - 2(x^2 - 6x + 9) = 24 + 3 \cdot 4 - 2 \cdot 9$$
$$3(y-2)^2 - 2(x-3)^2 = 18$$
$$\frac{(y-2)^2}{6} - \frac{(x-3)^2}{9} = 1$$

continued

The conic is a hyperbola with center at $(3, 2)$, axis parallel to the y-axis, and $a = \sqrt{6}$ and $b = 3$.

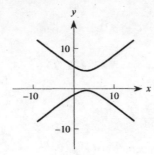

10. Since the foci are at $(-8, 0)$ and $(8, 0)$ and the vertices are at $(-10, 0)$ and $(10, 0)$, we know that the ellipse has major axis parallel to the x-axis with $a = 10$, $c = 8$, and $b = 6$. The center is at $(0, 0)$, so an equation for the ellipse is:

$$\frac{x^2}{100} + \frac{y^2}{36} = 1$$

11. Since the minor axis is 4, we know that $b = 2$. Since the center $(-6, 3)$ and the focus $(0, 3)$ are on the horizontal line $y = 3$, the major axis is parallel to the x-axis and $c = 6$. Thus, $a^2 = b^2 + c^2 = 2^2 + 6^2 = 40$ and an equation for the ellipse is:

$$\frac{(x+6)^2}{40} + \frac{(y-3)^2}{4} = 1$$

12. Since the center $(2, 3)$, the focus $(2, 7)$, and the corresponding vertex $(2, 6)$ all lie on the vertical line $x = 2$, the hyperbola's axis is parallel to the y-axis. From the given points we know that $a = 3$ and $c = 4$. Thus $b^2 = c^2 - a^2 = 4^2 - 3^2 = 7$, so an equation of the hyperbola is:

$$\frac{(y-3)^2}{9} + \frac{(x-2)^2}{7} = 1$$

13. Since the vertices are at $(0, 4)$ and $(0, -4)$, the center of the hyperbola is $(0, 0)$ with $a = 4$ and the axis is parallel to the y-axis. Thus, the equation is of the form

$$\frac{y^2}{a^2} - \frac{x^2}{b^2} = 1.$$

Since $(6, \sqrt{80})$ is a point on the hyperbola, it must satisfy its equation. Thus, we have:

$$\frac{80}{a^2} - \frac{36}{b^2} = 1$$
$$\frac{80}{16} - \frac{36}{b^2} = 1 \qquad \text{Substitute } a = 4.$$
$$b^2 = 9$$
$$b = 3$$

So the equation is $\dfrac{y^2}{16} - \dfrac{x^2}{9} = 1$.

14. The conic is an ellipse with foci at $(12, 0)$ and $(-12, 0)$. Hence $c = 12$, and with $2a = 30$, $a = 15$. We can find b from the equation   The equation of this ellipse is:

$$\frac{x^2}{15^2} - \frac{y^2}{9^2} = 1, \text{ or } \frac{x^2}{225} + \frac{y^2}{81} = 1$$

15. Since the foci are $F_1(-3, 0)$ and $F_2(3, 0)$, the ellipse has its major axis parallel to the x-axis and $c = 0$. Using the

geometric definition of an ellipse with $P\left(\dfrac{\sqrt{3}}{2}, 2\right) = (x, y)$, we get:

$$|F_1P| + |F_2P| = 2a \implies \sqrt{\left(\dfrac{\sqrt{3}}{2} - 3\right)^2 + (2-0)^2} + \sqrt{\left(\dfrac{\sqrt{3}}{2} + 3\right)^2 + (2-0)^2} = 2a \implies 7.277 = 2a$$

$$\sqrt{(x-c)^2 + y^2} + \sqrt{(x+c)^2 + y^2} = 2a$$

$$\sqrt{(x-3)^2 + y^2} + \sqrt{(x+3)^2 + y^2} = 7.277$$

$$\sqrt{x^2 - 6x + 9 + y^2} + \sqrt{x^2 + 6x + 9 + y^2} = 7.277$$

$$\sqrt{x^2 - 6x + 9 + y^2} = 7.277 - \sqrt{x^2 + 6x + 9 + y^2}$$

$$x^2 - 6x + 9 + y^2 = 7.277^2 - 2(7.277)\sqrt{x^2 + 6x + 9 + y^2} + x^2 + 6x + 9 + y^2$$

$$2(7.277)\sqrt{x^2 + 6x + 9 + y^2} = 7.277^2 + 12x$$

$$4(7.277)^2(x^2 + 6x + 9 + y^2) = 7.277^4 + 24(7.277)^2 x + 144x^2$$

$$x^2 + 6x + 9 + y^2 = \dfrac{7.277^2}{4} + 6x + \dfrac{36x^2}{7.277^2}$$

$$\dfrac{(7.277^2 - 36)x^2}{7.277^2} + y^2 = \dfrac{7.277^2 - 36}{4}$$

$$\dfrac{x^2}{3.1233014} + y^2 = 4.238682$$

$$\dfrac{x^2}{13.24} + \dfrac{y^2}{4.24} = 1$$

16. Visualizing the ceiling as the upper part of an ellipse, we get $b = 16$ and $a = 20$. So $c^2 = a^2 - b^2 = 400 - 256 = 144$, and $c = 12$. Thus, the two people should stand 12 feet on either side of the point below the highest point of the structure; that is, they should stand 24 feet apart along the vertices.

17. The given measurements specify $2c = 23$ and $a - c = 3$. Using $a^2 = b^2 - c^2$, we find that $a = 29/2$ and $b = \sqrt{78}$. This gives the equation

$$\dfrac{x^2}{\frac{841}{4}} + \dfrac{y^2}{78} = 1.$$

18. The points are $(6, -4)$, $(5, -1)$, $(4, 2)$, $(3, 5)$, $(2, 8)$. The function is $y = -3x + 14$.

19. The parametric curve is the same as the parabola $y = \dfrac{x^2}{9} + 1$ between $x = -6$ and $x = 6$.

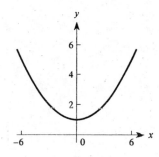

20. a.

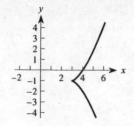

b.
$$y = t^3 - 1$$
$$y + 1 = t^3$$
$$(y+1)^{1/3} = t$$

Substituting $t = (y+1)^{1/3}$ in the equation $x = t^2 + 3$ gives $x = (y+1)^{2/3} + 3$.

c. When $y = 0$, $x = 4$.

21.

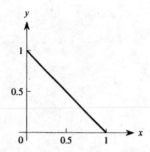

a.
$$x = 1 - \log t$$
$$\log t = 1 - x$$
$$10^{\log t} = 10^{1-x}$$
$$t = 10^{1-x}$$

Substitute $t = 10^{1-x}$: $y = \log t$
$$y = \log 10^{(1-x)}$$
$$y = 1 - x$$

b. The largest possible domain for this function is $0 \le x \le 1$.

22.

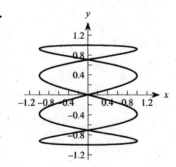

23. **a.**

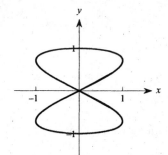

The graph has 2 "loops" and has period π.

b.

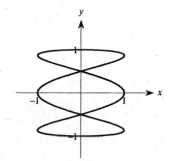

The graph has 3 "loops" and has period π.

c.

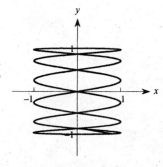

The graph has 6 "loops" and has period 2π.

24. $x = \cos 2t \Rightarrow x = 1 - 2\sin^2 t$. Since $y = \sin t$, we have $x = 1 - 2(\sin t)^2 \Rightarrow x = 1 - 2y^2$.

25. **a.** $\left. \begin{array}{l} r^2 = x^2 + y^2 \Rightarrow r^2 = 3^2 + 3^2 = 18 \Rightarrow r = \sqrt{18} \\ \theta = \arctan(3/3) = \arctan 1 \Rightarrow \theta = \pi/4 \end{array} \right\} \Rightarrow \left(\sqrt{18}, \dfrac{\pi}{4} \right)$

b. $\left. \begin{array}{l} r^2 = x^2 + y^2 \Rightarrow r^2 = (-1)^2 + 3^2 = 10 \Rightarrow r = \sqrt{10} \\ \theta = \arctan(3/-1) = \arctan(-3) \Rightarrow \theta = 1.89255 \end{array} \right\} \Rightarrow \left(\sqrt{10}, 1.89255 \right)$

c. $\left. \begin{array}{l} r^2 = x^2 + y^2 \Rightarrow r^2 = 4^2 + (-1)^2 = 17 \Rightarrow r = \sqrt{17} \\ \theta = \arctan(-1/4) \Rightarrow \theta = -0.244979 \end{array} \right\} \Rightarrow \left(\sqrt{17}, -0.244979 \right)$

d. $\left. \begin{array}{l} r^2 = x^2 + y^2 \Rightarrow r^2 = 0^2 + 6^2 = 36 \Rightarrow r = 6 \\ \theta = \arctan(6/0), \text{ which is undef.} \Rightarrow \theta = \pi/2 \end{array} \right\} \Rightarrow \left(6, \dfrac{\pi}{2} \right)$

26. **a.** $x = 3\cos \dfrac{\pi}{3} \Rightarrow x = 1.5, \quad y = 3\sin \dfrac{\pi}{3} \Rightarrow y = 1.5\sqrt{3} \Rightarrow (1.5, 1.5\sqrt{3})$

b. $x = 3\cos \dfrac{\pi}{4} \Rightarrow x = \dfrac{3\sqrt{2}}{2}, \quad y = 3\sin \dfrac{\pi}{4} \Rightarrow y = \dfrac{3\sqrt{2}}{2} \Rightarrow \left(\dfrac{3\sqrt{2}}{2}, \dfrac{3\sqrt{2}}{2} \right)$

c. $x = 4\cos \dfrac{3\pi}{2} \Rightarrow x = 0, \quad y = 4\sin \dfrac{3\pi}{2} \Rightarrow y = -4 \Rightarrow (0, -4)$

d. $x = 4\cos\frac{5\pi}{4} \Rightarrow x = -2\sqrt{2},\quad y = 4\sin\frac{5\pi}{4} \Rightarrow y = -2\sqrt{2} \Rightarrow (-2\sqrt{2}, -2\sqrt{2})$

e. $x = 5\cos\frac{5\pi}{6} \Rightarrow x = \frac{5\sqrt{3}}{2},\quad y = 5\sin\frac{5\pi}{6} \Rightarrow y = \frac{5}{2} \Rightarrow \left(\frac{5\sqrt{3}}{2}, \frac{5}{2}\right)$

f. $x = 5\cos 2 \Rightarrow x = -2.08073,\quad y = 5\sin 2 \Rightarrow y = 4.54649 \Rightarrow (-2.08073, 4.54649)$

27. The equation $r = \dfrac{1}{1+\cos\theta}$ can be rewritten as $r + r\cos\theta = 1$. Substituting x for $r\cos\theta$ and isolating the remaining r gives us $r = 1 - x$. Squaring both sides and substituting $r^2 = x^2 + y^2$ gives $x^2 + y^2 = 1 - 2x + x^2$. Rewriting this equation gives $x - \frac{1}{2} = \frac{1}{2}y^2$, which is a parabola whose vertex is $\left(\frac{1}{2}, 0\right)$ and opens to the left.

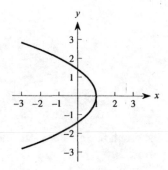

28.
$$r = \frac{1}{\sqrt{\dfrac{\cos^2\theta}{a^2} + \dfrac{\sin^2\theta}{b^2}}}$$

$$r^2 = \frac{1}{\dfrac{\cos^2\theta}{a^2} + \dfrac{\sin^2\theta}{b^2}}$$

$$\frac{r^2\cos^2\theta}{a^2} + \frac{r^2\sin^2\theta}{b^2} = 1 \qquad \text{Substitute } x = r\cos\theta \text{ and } y = r\sin\theta.$$

$$\frac{x^2}{a^2} + \frac{y^2}{b^2} = 1$$

This equation represents a family of ellipses.

29.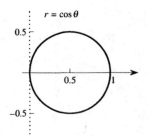

The graph is a circle along the polar axis.

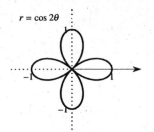

The graph is a 4-petaled rose along the axes.

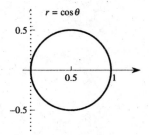

The graph is an 8-petaled rose along the axes.

30.

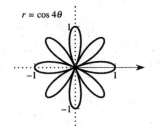

The graph is a circle along the polar axis.

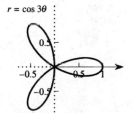

The graph is a 3-petaled rose with one petal along the polar axis.

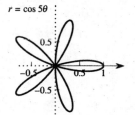

The graph is a 5-petaled rose with one petal along the polar axis.

31. $r = \sin\theta$

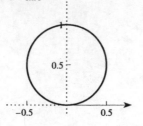

The graph is a circle along the positive vertical axis.

$r = \sin 2\theta$

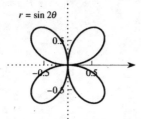

The graph is a 4-petaled rose between the axes.

$r = \sin 4\theta$

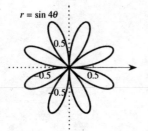

The graph is an 8-petaled rose between the axes.

32. $r = \sin\theta$

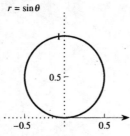

The graph is a circle along the positive vertical axis.

$r = \sin 3\theta$

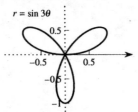

The graph is a 3-petaled rose with one petal along the negative vertical axis..

$r = \sin 5\theta$

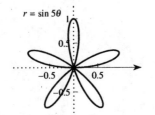

The graph is a 5-petaled rose with one petal along the positive vertical axis.

Chapter 10 Matrix Algebra and Its Applications

<u>Section 10.1 Geometric Vectors</u>

1.   **a.**

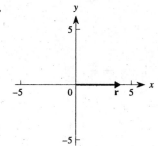

                                       b.

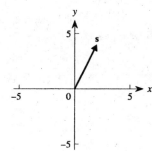

c.

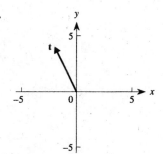

                                       d.

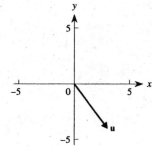

e.

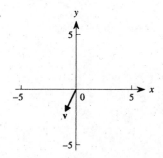

3.   **a.**
       b.
       c.

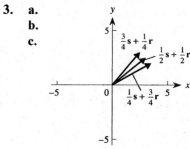

                                       d.

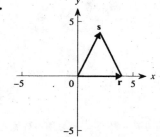

The final points of all the vectors drawn lie on this line.

5. **a.** $A = \begin{bmatrix} 1 & 2 \end{bmatrix}, \ B = \begin{bmatrix} 5 & 5 \end{bmatrix}$

$B - A = \begin{bmatrix} 5-1 & 5-2 \end{bmatrix} = \begin{bmatrix} 4 & 3 \end{bmatrix}$

$\| B - A \| = \sqrt{4^2 + 3^2} = 5$

b. $A = \begin{bmatrix} -2 & -1 \end{bmatrix}, \ B = \begin{bmatrix} 4 & 1 \end{bmatrix}$

$B - A = \begin{bmatrix} 4-(-2) & 1-(-1) \end{bmatrix} = \begin{bmatrix} 6 & 2 \end{bmatrix}$

$\| B - A \| = \sqrt{6^2 + 2^2} = \sqrt{40}$

c. $A = \begin{bmatrix} 1 & -3 \end{bmatrix}, \ B = \begin{bmatrix} -3 & -4 \end{bmatrix}$

$B - A = \begin{bmatrix} -3-1 & -4-(-3) \end{bmatrix} = \begin{bmatrix} -4 & -1 \end{bmatrix}$

$\| B - A \| = \sqrt{(-4)^2 + (-1)^2} = \sqrt{17}$

d. $A = \begin{bmatrix} 1 & 2 & 3 \end{bmatrix}, \ B = \begin{bmatrix} 4 & 5 & 3 \end{bmatrix}$

$B - A = \begin{bmatrix} 4-1 & 5-2 & 3-3 \end{bmatrix} = \begin{bmatrix} 3 & 3 & 0 \end{bmatrix}$

$\| B - A \| = \sqrt{3^2 + 3^2 + 0^2} = 3\sqrt{2}$

e. $A = \begin{bmatrix} -1 & 6 & 3 \end{bmatrix}, \ B = \begin{bmatrix} 3 & -1 & 2 \end{bmatrix}$

$B - A = \begin{bmatrix} 3-(-1) & -1-6 & 2-3 \end{bmatrix} = \begin{bmatrix} 4 & -7 & -1 \end{bmatrix}$

$\| B - A \| = \sqrt{4^2 + (-7)^2 + (-1)^2} = \sqrt{66}$

7. $\mathbf{v} = \begin{bmatrix} 3 & -1 & 2 \end{bmatrix}$

a. $3\mathbf{v} = \begin{bmatrix} 3 \cdot 3 & 3 \cdot (-1) & 3 \cdot 2 \end{bmatrix} = \begin{bmatrix} 9 & -3 & 6 \end{bmatrix}$

b. $10\mathbf{v} = \begin{bmatrix} 10 \cdot 3 & 10 \cdot (-1) & 10 \cdot 2 \end{bmatrix} = \begin{bmatrix} 30 & -10 & 20 \end{bmatrix}$

c. $-7\mathbf{v} = \begin{bmatrix} -7 \cdot 3 & -7 \cdot (-1) & -7 \cdot 2 \end{bmatrix} = \begin{bmatrix} -21 & 7 & -14 \end{bmatrix}$

d. $\frac{1}{2}\mathbf{v} = \begin{bmatrix} \frac{1}{2} \cdot 3 & \frac{1}{2} \cdot (-1) & \frac{1}{2} \cdot 2 \end{bmatrix} = \begin{bmatrix} \frac{3}{2} & -\frac{1}{2} & 1 \end{bmatrix}$

9. **a.** $\begin{bmatrix} 2 & 1 \end{bmatrix} = 2\mathbf{i} + \mathbf{j}$

b. $\begin{bmatrix} -1 & 3 \end{bmatrix} = -\mathbf{i} + 3\mathbf{j}$

c. $\begin{bmatrix} \frac{1}{2} & \frac{1}{2} \end{bmatrix} = \frac{1}{2}\mathbf{i} + \frac{1}{2}\mathbf{j}$

d. $\begin{bmatrix} 3 & 0 \end{bmatrix} = 3\mathbf{i}$

11.

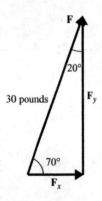

a. $\| \mathbf{F}_y \| = 30\sin(90° - 25°) = 30(0.9063) = 27.19$ pounds

b. $\| \mathbf{F}_y \| = 20\sin(90° - 30°) = 20(0.8660) = 17.32$ pounds

c. $\| \mathbf{F}_y \| = 40\sin(90° - 15°) = 40(0.2588) = 38.64$ pounds

13.

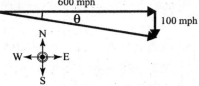

a. We choose a coordinate system with east along the positive x-axis and north along the positive y-axis. The plane's heading is $\mathbf{h} = \begin{bmatrix} 600 & 0 \end{bmatrix}$ and the wind's velocity is $\mathbf{w} = \begin{bmatrix} 0 & -100 \end{bmatrix}$. The plane's actual velocity is $\mathbf{h} + \mathbf{w} = \begin{bmatrix} 600 & -100 \end{bmatrix}$. To find the direction we note that $\tan\theta = \dfrac{100}{600}$, where θ is the angle south of east. Then $\theta = \arctan\left(\dfrac{100}{600}\right) = 9.46°$, so the direction is 9.46° south of east. The plane's actual speed is $\left\| \begin{bmatrix} 600 & -100 \end{bmatrix} \right\| = 608.28$ mph.

b. The new heading and wind vectors are $\mathbf{h} = \begin{bmatrix} 600 & 0 \end{bmatrix}$ and $\mathbf{w} = \begin{bmatrix} 100\cos 40° & -100\sin 40° \end{bmatrix}$; note that the second component of the wind vector is negative since the angle is 40° *south* of east. Calculating the components of $\mathbf{w}$ we find $\mathbf{w} = \begin{bmatrix} 76.60 & -64.28 \end{bmatrix}$. Thus the plane's actual velocity vector is

$$\mathbf{h} + \mathbf{w} = \begin{bmatrix} 600 & 0 \end{bmatrix} + \begin{bmatrix} 76.60 & -64.28 \end{bmatrix} = \begin{bmatrix} 676.60 & -64.28 \end{bmatrix}. \text{ Then } \theta = \arctan\left(\frac{64.28}{676.6}\right) = 5.43° \text{ south of}$$

east. The actual speed is $\left\| \begin{bmatrix} 676.6 & -64.28 \end{bmatrix} \right\| = 679.56$ mph.

c. In this scenario we need to compute components for both $\mathbf{h}$ and $\mathbf{w}$.
$$\mathbf{h} = \begin{bmatrix} -300\cos 45° & -300\sin 45° \end{bmatrix} = \begin{bmatrix} -212.13 & -212.13 \end{bmatrix} \text{ and}$$
$$\mathbf{w} = \begin{bmatrix} -100\cos 50° & -100\sin 50° \end{bmatrix} = \begin{bmatrix} -64.28 & -76.60 \end{bmatrix}. \text{ Then } \mathbf{w} + \mathbf{h} = \begin{bmatrix} -276.41 & -288.74 \end{bmatrix}. \text{ The}$$

plane's heading is $\theta = \arctan\left(\dfrac{288.74}{276.41}\right) = 46.25°$ south of west, and its actual speed is

$\left\| \begin{bmatrix} -276.41 & -288.74 \end{bmatrix} \right\| = 399.71$ mph. Note that we expect an answer close to 400 mph, since the plane is traveling almost exactly with the wind, so that the wind's speed is nearly added to the plane's speed.

Section 10.2 Linear Models

1. $\begin{bmatrix} 64 & 73 & 86 & 85 \\ 82 & 69 & 77 & 91 \\ 82 & 84 & 81 & 83 \end{bmatrix}$ with rows for Ted, Carol, and Alice, and columns for German, physics, theater, and politics.

3. a. HORN   **b.** MOTHER   **c.** EARLY   **d.** YESTERDAY

5. The production matrix is $\begin{bmatrix} 8 & 5 & 3 \\ 2 & 5 & 5 \\ 3 & 7 & 6 \end{bmatrix}$ with thousands of gallons of heating oil, diesel oil, and gasoline per shipment of crude in the rows and the three refineries in the columns. Let x_i be the number of shipments of crude processed by the ith refinery. Then $8x_1 + 5x_2 + 3x_3$ is the total output of heating oil from the three refineries, and we want this to equal 6200. Working through the other rows in the same way we get the system

$$8x_1 + 5x_2 + 3x_3 = 6200$$
$$2x_1 + 5x_2 + 5x_3 = 4000$$
$$3x_1 + 7x_2 + 6x_3 = 4700$$

7. Let x_1 be the number of microcomputers, x_2 be the number of terminals, and x_3 be the number of workstations. Then these three variables satisfy the following system of equations:

$$2000x_1 + 500x_2 + 5000x_3 = 280000$$
$$5x_1 - x_2 = 0$$
$$-x_1 + 2x_3 = 0$$

The first equation gives the total budget, the second relates terminals to microcomputers, and the third relates microcomputer to workstations.

9. The market transition matrix in the text is

$$\text{today}$$
$$\text{u} \quad \text{d} \quad \text{s}$$
$$\mathbf{A} = \begin{bmatrix} \frac{1}{4} & \frac{1}{2} & \frac{1}{4} \\ \frac{1}{2} & \frac{1}{4} & \frac{1}{2} \\ \frac{1}{4} & \frac{1}{4} & \frac{1}{4} \end{bmatrix} \begin{matrix} \text{up} \\ \text{down} \\ \text{same} \end{matrix} \text{tomorrow}$$

For each set of probabilities for today's market, p_1, p_2, and p_3, tomorrow's probabilities are given by

$$p_1^+ = \frac{1}{4}p_1 + \frac{1}{2}p_2 + \frac{1}{4}p_3$$
$$p_2^+ = \frac{1}{2}p_1 + \frac{1}{4}p_2 + \frac{1}{2}p_3$$
$$p_3^+ = \frac{1}{4}p_1 + \frac{1}{4}p_2 + \frac{1}{4}p_3$$

For each part of the problem, we will write today's probabilities as a row vector $\mathbf{p}$ and tomorrow's as a row vector $\mathbf{p}^+$.

a. $\mathbf{p} = \begin{bmatrix} 1 & 0 & 0 \end{bmatrix}$; $\mathbf{p}^+ = \begin{bmatrix} \frac{1}{4} & \frac{1}{2} & \frac{1}{4} \end{bmatrix}$

b. $\mathbf{p} = \begin{bmatrix} 0 & \frac{1}{2} & \frac{1}{2} \end{bmatrix}$; $\mathbf{p}^+ = \begin{bmatrix} \frac{3}{8} & \frac{3}{8} & \frac{1}{4} \end{bmatrix}$

c. $\mathbf{p} = \begin{bmatrix} \frac{1}{2} & 0 & \frac{1}{2} \end{bmatrix}$; $\mathbf{p}^+ = \begin{bmatrix} \frac{1}{4} & \frac{1}{2} & \frac{1}{4} \end{bmatrix}$

d. $\mathbf{p} = \begin{bmatrix} \frac{1}{4} & \frac{1}{2} & \frac{1}{4} \end{bmatrix}$; $\mathbf{p}^+ = \begin{bmatrix} \frac{3}{8} & \frac{3}{8} & \frac{1}{4} \end{bmatrix}$

e. $\mathbf{p} = \begin{bmatrix} 0.35 & 0.40 & 0.25 \end{bmatrix}$; $\mathbf{p}^+ = \begin{bmatrix} 0.35 & 0.40 & 0.25 \end{bmatrix}$

11.　**a.** The matrix of transition probabilities is

this week

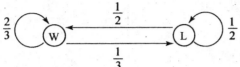

$$\begin{array}{cc} \text{W} & \text{L} \end{array}$$
$$\begin{bmatrix} \frac{2}{3} & \frac{1}{2} \\ \frac{1}{3} & \frac{1}{2} \end{bmatrix} \begin{array}{c} \text{W} \\ \text{L} \end{array} \quad \text{next week}$$

b. The probabilities for this week are $\frac{1}{2}$ win and $\frac{1}{2}$ lose. The chance of winning next week's game is

$$\frac{2}{3}\cdot\frac{1}{2}+\frac{1}{2}\cdot\frac{1}{2}=\frac{7}{12}.$$

c. If the team win's today, next week's probabilities are

Win: $\quad 1\cdot\frac{2}{3}=\frac{2}{3}$

Lose: $\quad 1-\frac{2}{3}=\frac{1}{3}$

Then the probability of winning the week after is $\dfrac{2}{3}\cdot\dfrac{2}{3}+\dfrac{1}{2}\cdot\dfrac{1}{3}=\dfrac{11}{18}.$

13.　**a.** With I, E, S, and M standing for ignorance, exploratory thinking, superficial understanding, and mastery, the Markov chain is defined by the matrix

now

$$\begin{array}{cccc} \text{I} & \text{E} & \text{S} & \text{M} \end{array}$$
$$\mathbf{A} = \begin{bmatrix} \frac{1}{2} & \frac{1}{4} & 0 & 0 \\ \frac{1}{2} & \frac{1}{2} & \frac{1}{4} & 0 \\ 0 & \frac{1}{4} & \frac{1}{2} & 0 \\ 0 & 0 & \frac{1}{4} & 1 \end{bmatrix} \begin{array}{c} \text{I} \\ \text{E} \\ \text{S} \\ \text{M} \end{array} \quad \text{after one lesson}$$

b. State I corresponds to the starting probability vector $\begin{bmatrix} 1 \\ 0 \\ 0 \\ 0 \end{bmatrix}$. After one lesson your vector is $\begin{bmatrix} \frac{1}{2} \\ \frac{1}{2} \\ 0 \\ 0 \end{bmatrix}$ where

we show only the products corresponding to the first column of **A**. Since only the first two coordinates of this result are nonzero, we need only the first two columns of **A** to compute the result after two

lessons. It is $\begin{bmatrix} \left(\frac{1}{2}\right)\left(\frac{1}{2}\right)+\left(\frac{1}{4}\right)\left(\frac{1}{2}\right) \\ \left(\frac{1}{2}\right)\left(\frac{1}{2}\right)+\left(\frac{1}{2}\right)\left(\frac{1}{2}\right) \\ (0)\left(\frac{1}{2}\right)+\left(\frac{1}{4}\right)\left(\frac{1}{2}\right) \\ (0)\left(\frac{1}{2}\right)+(0)\left(\frac{1}{2}\right) \end{bmatrix} = \begin{bmatrix} \frac{3}{8} \\ \frac{1}{2} \\ \frac{1}{8} \\ 0 \end{bmatrix}$. After three days: $\begin{bmatrix} \left(\frac{1}{2}\right)\left(\frac{3}{8}\right)+\left(\frac{1}{4}\right)\left(\frac{1}{2}\right)+(0)\left(\frac{1}{8}\right) \\ \left(\frac{1}{2}\right)\left(\frac{3}{8}\right)+\left(\frac{1}{2}\right)\left(\frac{1}{2}\right)+\left(\frac{1}{4}\right)\left(\frac{1}{8}\right) \\ (0)\left(\frac{3}{8}\right)+\left(\frac{1}{4}\right)\left(\frac{1}{2}\right)+\left(\frac{1}{2}\right)\left(\frac{1}{8}\right) \\ (0)\left(\frac{3}{8}\right)+(0)\left(\frac{1}{2}\right)+\left(\frac{1}{4}\right)\left(\frac{1}{8}\right) \end{bmatrix} = \begin{bmatrix} \frac{5}{16} \\ \frac{15}{32} \\ \frac{3}{16} \\ \frac{1}{32} \end{bmatrix}.$

Note, as a useful check, that for each day the vector of probabilities must add to 1.

15. a. The equations for the model are

$$C^+ = 1.2C - 0.3S$$

$$S^+ = -0.2C + 1.2S$$

We will assume that the time unit is months, so that C^+ and S^+ are the numbers of cattle and sheep after one month when C and S are the initial populations. Computing as in Example 3, we find

	Cattle	Sheep
Start	50	100
1 month	30	110
2 months	3	126
3 months	−34.2	150.6

Note that cattle will disappear during the third month. According to the model, in the absence of cattle, the sheep population grows by 20% per month, so the sheep population will grow out of control.

b. The equations for the model are

$$C^+ = 1.2C - 0.1S$$

$$S^+ = 0.5C + 1.4S$$

Computing as in Example 3 we find

	Cattle	Sheep
Start	50	100
1 month	50	165
2 months	43.5	256
3 months	26.6	380.15

17. a. We need to solve the system

$$H = 1.2H = 0.3W$$

$$W = 0.2H + 0.7W$$

From the first equation, $0.3W = 0.2H$. The second equation yields the same information, so any pair of positive values H and W for which $W = \frac{2}{3}H$ is a possible solution.

b. If $(H_0, W_0) = (1000, 800)$, the next point is $\left((1.2)(1000) - (0.3)(800),\ (0.2)(1000) + (0.7)(800)\right)$ or

(960, 760). The slope of this line is 1 and its equation is $\dfrac{800 - W}{1000 - H} = 1$ or $W = H - 200$.

c. To find the limiting population values, we solve the equations for the trajectory and the limiting line as a system. If the solutions are positive, they represent the limiting populations. If either solution is negative, we must examine what happens when the corresponding species dies out. For $(H_0, W_0) = (1000, 800)$, we have the system

$$W = H - 200$$

$$W = \frac{2}{3}H$$

which has the solution $(H, W) = (600, 400)$, so the limiting population is 600 groups of hares and 400 wolves.

Section 10.3 Scalar Products

1. $a = \begin{bmatrix} 3 \\ 5 \end{bmatrix}$, $b = \begin{bmatrix} 0 \\ 2 \end{bmatrix}$, $c = \begin{bmatrix} 5 \\ -1 \end{bmatrix}$, $d = \begin{bmatrix} -1 \\ 0 \end{bmatrix}$, $e = \begin{bmatrix} 4 \\ 5 \end{bmatrix}$

  a. $a \cdot c = 3(5) + 5(-1) = 10$

  b. $b \cdot c = 0(5) + 2(-1) = -2$

  c. $b \cdot d = 0(-1) + 2(0) = 0$

  d. $a \cdot d = 3(-1) + 5(0) = -3$

  e. $c \cdot d = 5(-1) + (-1)(0) = -5$

  f. $b \cdot e = 0(4) + 2(5) = 10$

  g. $c \cdot e = 5(4) + (-1)(5) = 15$

3. $a = \begin{bmatrix} 2 \\ 5 \\ 1 \end{bmatrix}$, $b = \begin{bmatrix} 0 \\ 1 \\ 0 \end{bmatrix}$, $c = \begin{bmatrix} 7 \\ -2 \\ -1 \end{bmatrix}$, $d = \begin{bmatrix} -2 \\ 3 \\ 1 \end{bmatrix}$

  a. $a \cdot c = 2(7) + 5(-2) + 1(-1) = 3$

  b. $b \cdot c = 0(7) + 1(-2) + 0(-1) = -2$

  c. $b \cdot d = 0(-2) + 1(3) + 0(1) = 3$

  d. $a \cdot d = 2(-2) + 5(3) + 1(1) = 12$

  e. $a \cdot a = 2(2) + 5(5) + 1(1) = 30$

  f. $c \cdot d = 7(-2) + (-2)(3) + (-1)(1) = -21$

5. $a = \begin{bmatrix} 3 \\ 5 \end{bmatrix}$, $b = \begin{bmatrix} 0 \\ 2 \end{bmatrix}$, $c = \begin{bmatrix} 5 \\ -1 \end{bmatrix}$, $A = \begin{bmatrix} 1 & 7 \\ 4 & 2 \end{bmatrix}$, $B = \begin{bmatrix} 5 & 2 \\ -1 & 0 \end{bmatrix}$, $C = \begin{bmatrix} 2 & 1 \\ 4 & -2 \\ 3 & 1 \end{bmatrix}$

  a. $Ab = \begin{bmatrix} 1 & 7 \\ 4 & 2 \end{bmatrix} \begin{bmatrix} 0 \\ 2 \end{bmatrix} = \begin{bmatrix} 1(0) + 7(2) \\ 4(0) + 2(2) \end{bmatrix} = \begin{bmatrix} 14 \\ 4 \end{bmatrix}$

  b. $Ac = \begin{bmatrix} 1 & 7 \\ 4 & 2 \end{bmatrix} \begin{bmatrix} 5 \\ -1 \end{bmatrix} = \begin{bmatrix} 1(5) + 7(-1) \\ 4(5) + 2(-1) \end{bmatrix} = \begin{bmatrix} -2 \\ 18 \end{bmatrix}$

  c. $Ba = \begin{bmatrix} 5 & 2 \\ -1 & 0 \end{bmatrix} \begin{bmatrix} 3 \\ 5 \end{bmatrix} = \begin{bmatrix} 5(3) + 2(5) \\ (-1)(3) + 0(5) \end{bmatrix} = \begin{bmatrix} 25 \\ -3 \end{bmatrix}$

  d. $Bb = \begin{bmatrix} 5 & 2 \\ -1 & 0 \end{bmatrix} \begin{bmatrix} 0 \\ 2 \end{bmatrix} = \begin{bmatrix} 5(0) + 2(2) \\ (-1)(0) + 0(2) \end{bmatrix} = \begin{bmatrix} 4 \\ 0 \end{bmatrix}$

  e. $Cc = \begin{bmatrix} 2 & 1 \\ 4 & -2 \\ 3 & 1 \end{bmatrix} \begin{bmatrix} 5 \\ -1 \end{bmatrix} = \begin{bmatrix} 2(5) + 1(-1) \\ 4(5) + (-2)(-1) \\ 3(5) + 1(-1) \end{bmatrix} = \begin{bmatrix} 9 \\ 22 \\ 14 \end{bmatrix}$

  f. $Ca = \begin{bmatrix} 2 & 1 \\ 4 & -2 \\ 3 & 1 \end{bmatrix} \begin{bmatrix} 3 \\ 5 \end{bmatrix} = \begin{bmatrix} 2(3) + 1(5) \\ 4(3) + (-2)(5) \\ 3(3) + 1(5) \end{bmatrix} = \begin{bmatrix} 11 \\ 2 \\ 14 \end{bmatrix}$

7. $\mathbf{a} = \begin{bmatrix} 2 \\ 5 \\ 1 \end{bmatrix}$, $\mathbf{b} = \begin{bmatrix} 0 \\ 1 \\ 0 \end{bmatrix}$, $\mathbf{c} = \begin{bmatrix} 7 \\ -2 \\ -1 \end{bmatrix}$, $\mathbf{d} = \begin{bmatrix} -2 \\ 3 \\ 1 \end{bmatrix}$, $\mathbf{A} = \begin{bmatrix} 1 & 0 & 0 \\ 0 & 1 & 0 \\ 0 & 0 & 1 \end{bmatrix}$, $\mathbf{B} = \begin{bmatrix} 0 & 0 & 1 \\ 0 & 1 & 0 \\ 1 & 0 & 0 \end{bmatrix}$, $\mathbf{C} = \begin{bmatrix} 1 & 0 & 0 \\ 0 & -1 & 0 \\ 0 & 0 & 2 \end{bmatrix}$

a. $\mathbf{Ac} = \begin{bmatrix} 1 & 0 & 0 \\ 0 & 1 & 0 \\ 0 & 0 & 1 \end{bmatrix} \begin{bmatrix} 7 \\ -2 \\ -1 \end{bmatrix} = \begin{bmatrix} 1(7)+0(-2)+0(-1) \\ 0(7)+1(-2)+0(-1) \\ 0(7)+0(-2)+1(-1) \end{bmatrix} = \begin{bmatrix} 7 \\ -2 \\ -1 \end{bmatrix} = \mathbf{c}$

b. $\mathbf{Ad} = \begin{bmatrix} 1 & 0 & 0 \\ 0 & 1 & 0 \\ 0 & 0 & 1 \end{bmatrix} \begin{bmatrix} -2 \\ 3 \\ 1 \end{bmatrix} = \begin{bmatrix} 1(-2)+0(3)+0(1) \\ 0(-2)+1(3)+0(1) \\ 0(-2)+0(3)+1(1) \end{bmatrix} = \begin{bmatrix} -2 \\ 3 \\ 1 \end{bmatrix} = \mathbf{d}$

c. $\mathbf{Ba} = \begin{bmatrix} 0 & 0 & 1 \\ 0 & 1 & 0 \\ 1 & 0 & 0 \end{bmatrix} \begin{bmatrix} 2 \\ 5 \\ 1 \end{bmatrix} = \begin{bmatrix} 0(2)+0(5)+0(1) \\ 0(2)+1(5)+0(1) \\ 1(2)+0(5)+1(1) \end{bmatrix} = \begin{bmatrix} 1 \\ 5 \\ 2 \end{bmatrix}$

d. $\mathbf{Bd} = \begin{bmatrix} 0 & 0 & 1 \\ 0 & 1 & 0 \\ 1 & 0 & 0 \end{bmatrix} \begin{bmatrix} -2 \\ 3 \\ 1 \end{bmatrix} = \begin{bmatrix} 0(-2)+0(3)+0(1) \\ 0(-2)+1(3)+0(1) \\ 1(-2)+0(3)+1(1) \end{bmatrix} = \begin{bmatrix} 1 \\ 3 \\ -2 \end{bmatrix}$

e. $\mathbf{Cb} = \begin{bmatrix} 1 & 0 & 0 \\ 0 & -1 & 0 \\ 0 & 0 & 2 \end{bmatrix} \begin{bmatrix} 0 \\ 1 \\ 0 \end{bmatrix} = \begin{bmatrix} 1(0)+0(1)+0(0) \\ 0(0)+(-1)(1)+0(0) \\ 0(0)+0(1)+2(0) \end{bmatrix} = \begin{bmatrix} 0 \\ -1 \\ 0 \end{bmatrix}$

f. $\mathbf{Ca} = \begin{bmatrix} 1 & 0 & 0 \\ 0 & -1 & 0 \\ 0 & 0 & 2 \end{bmatrix} \begin{bmatrix} 2 \\ 5 \\ 1 \end{bmatrix} = \begin{bmatrix} 1(2)+0(5)+0(1) \\ 0(2)+(-1)(5)+0(1) \\ 0(2)+0(5)+2(1) \end{bmatrix} = \begin{bmatrix} 2 \\ -5 \\ 2 \end{bmatrix}$

9. a. With $\mathbf{A} = \begin{bmatrix} 5 & 1 \\ 4 & 3 \end{bmatrix}$ and $\mathbf{b} = \begin{bmatrix} 2 \\ 5 \end{bmatrix}$, $\mathbf{Ax} = \mathbf{b}$ corresponds to the system $\begin{aligned} 5x_1 + x_2 &= 2 \\ 4x_1 + 3x_2 &= 5 \end{aligned}$.

b. With $\mathbf{A} = \begin{bmatrix} 1 & 4 \\ 2 & -3 \end{bmatrix}$ and $\mathbf{b} = \begin{bmatrix} 4 \\ 9 \end{bmatrix}$, $\mathbf{Ax} = \mathbf{b}$ corresponds to the system $\begin{aligned} x_1 + 4x_2 &= 4 \\ 2x_1 - 3x_2 &= 9 \end{aligned}$.

c. With $\mathbf{A} = \begin{bmatrix} 5 & 2 & 1 \\ 4 & 1 & 6 \\ 3 & 1 & 0 \end{bmatrix}$ and $\mathbf{b} = \begin{bmatrix} 1 \\ 5 \\ 2 \end{bmatrix}$, $\mathbf{Ax} = \mathbf{b}$ corresponds to the system $\begin{aligned} 5x_1 + 2x_2 + x_3 &= 1 \\ 4x_1 + x_2 + 6x_3 &= 5 \\ 3x_1 + x_2 &= 2 \end{aligned}$.

d. With $\mathbf{A} = \begin{bmatrix} 2 & -1 & 5 \\ 3 & 1 & 2 \\ 5 & 1 & -3 \end{bmatrix}$, $\mathbf{b} = \begin{bmatrix} 0 \\ 0 \\ 0 \end{bmatrix}$, $\mathbf{Ax} = \mathbf{b}$ corresponds to the system $\begin{aligned} 2x_1 - x_2 + 5x_3 &= 0 \\ 3x_1 + x_2 + 2x_3 &= 0 \\ 5x_1 + x_2 - 3x_3 &= 0 \end{aligned}$.

11. Each system has the form $\mathbf{Ax} = \mathbf{b}$.

a. $\mathbf{A} = \begin{bmatrix} 6 & 4 & 2 \\ 4 & 8 & 4 \\ 3 & 2 & 8 \end{bmatrix}$, $\mathbf{b} = \begin{bmatrix} 400 \\ 800 \\ 500 \end{bmatrix}$

b. $A = \begin{bmatrix} 8 & 5 & 3 \\ 2 & 5 & 5 \\ 3 & 7 & 6 \end{bmatrix}$, $b = \begin{bmatrix} 6200 \\ 4000 \\ 4700 \end{bmatrix}$

c. $A = \begin{bmatrix} 10 & 50 & 200 \\ 1 & 3 & 0.2 \\ 30 & 10 & 0 \end{bmatrix}$, $b = \begin{bmatrix} 600 \\ 20 \\ 200 \end{bmatrix}$

d. $A = \begin{bmatrix} 2000 & 500 & 5000 \\ 5 & -1 & 0 \\ -1 & 0 & 2 \end{bmatrix}$, $b = \begin{bmatrix} 280,000 \\ 0 \\ 0 \end{bmatrix}$

13. a. The cost vector is **Ad**, where $A = \begin{bmatrix} 4 & 2 & 1.5 & 6 \\ 6 & 1 & 2 & 5 \\ 5 & 0.85 & 2.5 & 7 \end{bmatrix}$ and $d = \begin{bmatrix} 10 \\ 6 \\ 3 \\ 2 \end{bmatrix}$.

b. Caterer A, \$68.50; Caterer B, \$82; Caterer C, \$76.60.

15. a.

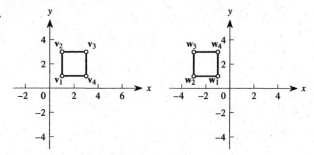

b.

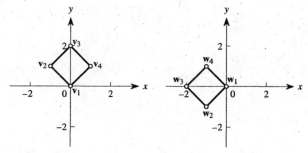

17. $w = Rv = \begin{bmatrix} \cos\theta & -\sin\theta \\ \sin\theta & \cos\theta \end{bmatrix} \begin{bmatrix} a \\ b \end{bmatrix} = \begin{bmatrix} a\cos\theta - b\sin\theta \\ a\sin\theta + b\cos\theta \end{bmatrix}$

$\| w \| = \sqrt{(a\cos\theta - b\sin\theta)^2 + (a\sin\theta + b\cos\theta)^2}$

$= \sqrt{\begin{aligned} & a^2\cos^2\theta - 2ab\cos\theta\sin\theta + b^2\sin^2\theta \\ & + a^2\sin^2\theta + 2ab\sin\theta\cos\theta + b^2\cos^2\theta \end{aligned}}$

$= \sqrt{(a^2 + b^2)(\cos^2\theta + \sin^2\theta)} = \sqrt{a^2 + b^2} = \| v \|$

19. a. Rotating clockwise through an angle θ is equivalent to rotating counterclockwise through an angle of $-\theta$, so the required matrix is

$$\begin{bmatrix} \cos(-\theta) & -\sin(-\theta) \\ \sin(-\theta) & \cos(-\theta) \end{bmatrix}.$$

Since $\cos(-\theta) = \cos\theta$ and $\sin(-\theta) = -\sin\theta$, this matrix is the same as

$$\begin{bmatrix} \cos\theta & \sin\theta \\ -\sin\theta & \cos\theta \end{bmatrix}.$$

b. $\begin{bmatrix} \cos 30° & -\sin 30° \\ \sin 30° & \cos 30° \end{bmatrix} = \begin{bmatrix} \dfrac{\sqrt{3}}{2} & -\dfrac{1}{2} \\ \dfrac{1}{2} & \dfrac{\sqrt{3}}{2} \end{bmatrix}$

c.

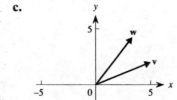

d. The multiplication produces the original vector $\mathbf{v} = \begin{bmatrix} 5 \\ 2 \end{bmatrix}$.

21. In each calculation, θ is the acute angle between the two vectors.

a. $\mathbf{a} = \begin{bmatrix} 3 & 5 \end{bmatrix}$, $\mathbf{b} = \begin{bmatrix} -2 & 4 \end{bmatrix}$

$$\cos\theta = \frac{\mathbf{a}\cdot\mathbf{b}}{\sqrt{\mathbf{a}\cdot\mathbf{a}}\sqrt{\mathbf{b}\cdot\mathbf{b}}} = \frac{3(-2)+5(4)}{\sqrt{3^2+5^2}\sqrt{(-2)^2+4^2}} \approx 0.53688$$

$$\theta = \arccos(0.53688) = 57.53°$$

b. $\mathbf{a} = \begin{bmatrix} 1 & 4 \end{bmatrix}$, $\mathbf{b} = \begin{bmatrix} -2 & 5 \end{bmatrix}$

$$\cos\theta = \frac{\mathbf{a}\cdot\mathbf{b}}{\sqrt{\mathbf{a}\cdot\mathbf{a}}\sqrt{\mathbf{b}\cdot\mathbf{b}}} = \frac{1(-2)+4(5)}{\sqrt{1^2+4^2}\sqrt{(-2)^2+5^2}} \approx 0.81068$$

$$\theta = \arccos(0.81068) = 35.84°$$

c. $\mathbf{a} = \begin{bmatrix} 1 & 4 & 5 \end{bmatrix}$, $\mathbf{b} = \begin{bmatrix} 2 & 3 & -2 \end{bmatrix}$

$$\cos\theta = \frac{\mathbf{a}\cdot\mathbf{b}}{\sqrt{\mathbf{a}\cdot\mathbf{a}}\sqrt{\mathbf{b}\cdot\mathbf{b}}} = \frac{1(2)+4(3)+5(-2)}{\sqrt{1^2+4^2+5^2}\sqrt{2^2+3^2+(-2)^2}} \approx 0.14970$$

$$\theta = \arccos(0.14970) = 81.39°$$

d. $\mathbf{a} = \begin{bmatrix} 6 & 4 & -1 \end{bmatrix}$, $\mathbf{b} = \begin{bmatrix} 5 & -3 & 2 \end{bmatrix}$

$$\cos\theta = \frac{\mathbf{a}\cdot\mathbf{b}}{\sqrt{\mathbf{a}\cdot\mathbf{a}}\sqrt{\mathbf{b}\cdot\mathbf{b}}} = \frac{6(5)+4(-3)+(-1)(2)}{\sqrt{6^2+4^2+(-1)^2}\sqrt{5^2+(-3)^2+2^2}} \approx 0.35653$$

$$\theta = \arccos(0.35653) = 69.11°$$

23. The vectors for Ivy, State and Hawaii are $\mathbf{I} = \begin{bmatrix} 8 & 9 & 7 & 7 & 5 \end{bmatrix}$, $\mathbf{S} = \begin{bmatrix} 7 & 6 & 9 & 6 & 7 \end{bmatrix}$,

    $\mathbf{H} = \begin{bmatrix} 10 & 7 & 6 & 4 & 6 \end{bmatrix}$. Let θ_{IS} be the angle between the vectors for Ivy and State, and let θ_{IH} be the angle between the vectors for Ivy and Hawaii.

    $$\cos\theta_{IS} = \frac{\mathbf{I}\cdot\mathbf{S}}{\sqrt{\mathbf{I}\cdot\mathbf{I}}\sqrt{\mathbf{S}\cdot\mathbf{S}}} = 0.96391; \quad \theta_{IS} = 15.44°$$

    $$\cos\theta_{IH} = \frac{\mathbf{I}\cdot\mathbf{H}}{\sqrt{\mathbf{I}\cdot\mathbf{I}}\sqrt{\mathbf{H}\cdot\mathbf{H}}} = 0.96419; \quad \theta_{IH} = 15.38°$$

    The Ivy-State angle and the Ivy-Hawaii angle are almost exactly equal. Susan might want to look for some other way of making her choice. (See Problem 24.)

25.

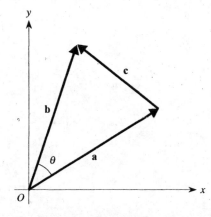

    Using $\mathbf{c} = \mathbf{b} - \mathbf{a}$ to compute $\mathbf{c}\cdot\mathbf{c}$ we find $\mathbf{c}\cdot\mathbf{c} = (\mathbf{b}-\mathbf{a})\cdot(\mathbf{b}-\mathbf{a}) = \mathbf{b}\cdot\mathbf{b} - 2(\mathbf{a}\cdot\mathbf{b}) + \mathbf{a}\cdot\mathbf{a}$ which we can write as

    $\|\mathbf{c}\|^2 = \|\mathbf{a}\|^2 + \|\mathbf{b}\|^2 - 2(\mathbf{a}\cdot\mathbf{b})$. According to the law of cosines, $\|\mathbf{c}\|^2 = \|\mathbf{a}\|^2 + \|\mathbf{b}\|^2 - 2\|\mathbf{a}\|\|\mathbf{b}\|\cos\theta$.

    Comparing these two expressions for $\|\mathbf{c}\|^2$ shows that $-2(\mathbf{a}\cdot\mathbf{b}) = -2\|\mathbf{a}\|\|\mathbf{b}\|\cos\theta$ and thus

    $$\cos\theta = \frac{\mathbf{a}\cdot\mathbf{b}}{\|\mathbf{a}\|\|\mathbf{b}\|}.$$

Section 10.4 Matrix Multiplication

1. $A = \begin{bmatrix} 0 & 2 \\ 1 & 4 \end{bmatrix}$, $B = \begin{bmatrix} 4 & 2 \\ 1 & 1 \end{bmatrix}$, $C = \begin{bmatrix} 3 & 1 \\ -1 & -3 \end{bmatrix}$

    a. $AB = \begin{bmatrix} 0(4)+2(1) & 0(2)+2(1) \\ 1(4)+4(1) & 1(2)+4(1) \end{bmatrix} = \begin{bmatrix} 2 & 2 \\ 8 & 6 \end{bmatrix}$

    b. $BA = \begin{bmatrix} 4(0)+2(1) & 4(2)+2(4) \\ 1(0)+1(1) & 1(2)+1(4) \end{bmatrix} = \begin{bmatrix} 2 & 16 \\ 1 & 6 \end{bmatrix}$

    c. $CB = \begin{bmatrix} 3(4)+1(1) & 3(2)+1(1) \\ (-1)(4)+(-3)(1) & (-1)(2)+(-3)(1) \end{bmatrix} = \begin{bmatrix} 13 & 7 \\ -7 & -5 \end{bmatrix}$

    d. $BC = \begin{bmatrix} 4(3)+2(-1) & 4(1)+2(-3) \\ 1(3)+1(-1) & 1(1)+1(-3) \end{bmatrix} = \begin{bmatrix} 10 & -2 \\ 2 & -2 \end{bmatrix}$

    e. $AC = \begin{bmatrix} 0(3)+2(-1) & 0(1)+2(-3) \\ 1(3)+4(-1) & 1(1)+4(-3) \end{bmatrix} = \begin{bmatrix} -2 & -6 \\ -1 & -11 \end{bmatrix}$

f. $A^2 = \begin{bmatrix} 0(0)+2(1) & 0(2)+2(4) \\ 1(0)+4(1) & 1(2)+4(4) \end{bmatrix} = \begin{bmatrix} 2 & 8 \\ 4 & 18 \end{bmatrix}$

g. $B^2 = \begin{bmatrix} 4(4)+2(1) & 4(2)+2(1) \\ 1(4)+1(1) & 1(2)+1(1) \end{bmatrix} = \begin{bmatrix} 18 & 10 \\ 5 & 3 \end{bmatrix}$

3. $A = \begin{bmatrix} 5 & 0 \\ 1 & 4 \end{bmatrix}$, $B = \begin{bmatrix} 5 & -1 \\ 0 & 2 \end{bmatrix}$, $C = \begin{bmatrix} 1 & 2 \\ 3 & 4 \\ 5 & 6 \end{bmatrix}$, $D = \begin{bmatrix} 1 & 5 & -2 \\ 3 & 0 & 2 \end{bmatrix}$

a. $AB = \begin{bmatrix} 5(5)+0(0) & 5(-1)+0(2) \\ 1(5)+4(0) & 1(-1)+4(2) \end{bmatrix} = \begin{bmatrix} 25 & -5 \\ 5 & 7 \end{bmatrix}$

b. $CB = \begin{bmatrix} 1(5)+2(0) & 1(-1)+2(2) \\ 3(5)+4(0) & 3(-1)+4(2) \\ 5(5)+6(0) & 5(-1)+6(2) \end{bmatrix} = \begin{bmatrix} 5 & 3 \\ 15 & 5 \\ 25 & 7 \end{bmatrix}$

c. The product **BC** is not defined.

d. $AD = \begin{bmatrix} 5(1)+0(3) & 5(5)+0(0) & 5(-2)+0(2) \\ 1(1)+4(3) & 1(5)+4(0) & 1(-2)+4(2) \end{bmatrix} = \begin{bmatrix} 5 & 25 & -10 \\ 13 & 5 & 6 \end{bmatrix}$

e. The product **DA** is not defined.

f. $CD = \begin{bmatrix} 1(1)+2(3) & 1(5)+2(0) & 1(-2)+2(2) \\ 3(1)+4(3) & 3(5)+4(0) & 3(-2)+4(2) \\ 5(1)+6(3) & 5(5)+6(0) & 5(-2)+6(2) \end{bmatrix} = \begin{bmatrix} 7 & 5 & 2 \\ 15 & 15 & 2 \\ 23 & 25 & 2 \end{bmatrix}$

g. $DC = \begin{bmatrix} 1(1)+5(3)+(-2)(5) & 1(2)+5(4)+(-2)(6) \\ 3(1)+0(3)+2(5) & 3(2)+0(4)+2(6) \end{bmatrix} = \begin{bmatrix} 6 & 10 \\ 13 & 18 \end{bmatrix}$

h. The product C^2 is not defined.

5. $B = \begin{bmatrix} 5 & -1 \\ 0 & 2 \end{bmatrix}$, $C = \begin{bmatrix} 1 & 2 \\ 3 & 4 \\ 5 & 6 \end{bmatrix}$, $D = \begin{bmatrix} 1 & 5 & -2 \\ 3 & 0 & 2 \end{bmatrix}$

a. $CB = \begin{bmatrix} 5 & 3 \\ 15 & 5 \\ 25 & 7 \end{bmatrix}$, $(CB)D = \begin{bmatrix} 14 & 25 & -4 \\ 30 & 75 & -20 \\ 46 & 125 & -36 \end{bmatrix}$

b. $BD = \begin{bmatrix} 2 & 25 & -12 \\ 6 & 0 & 4 \end{bmatrix}$, $C(BD) = \begin{bmatrix} 14 & 25 & -4 \\ 30 & 75 & -20 \\ 46 & 125 & -36 \end{bmatrix}$

In this example, $(CB)D = C(BD)$.

7. $A = \begin{bmatrix} -1 & 3 & -2 \\ 3 & 4 & -1 \\ 4 & 0 & 1 \\ 4 & 0 & 1 \end{bmatrix}$, $P = \begin{bmatrix} 0 & 0 & 1 \\ 0 & 1 & 0 \\ 1 & 0 & 0 \end{bmatrix}$, $Q = \begin{bmatrix} 0 & 1 & 0 \\ 0 & 0 & 1 \\ 1 & 0 & 0 \end{bmatrix}$, $R = \begin{bmatrix} 1 & 0 & 0 \\ 0 & 3 & 0 \\ 0 & 0 & 2 \end{bmatrix}$

a. $AP = \begin{bmatrix} -2 & 3 & -1 \\ -1 & 4 & 3 \\ 1 & 0 & 4 \\ 1 & 0 & 4 \end{bmatrix}$ This product switches the first and last columns of **A**.

b. $AQ = \begin{bmatrix} -2 & -1 & 3 \\ -1 & 3 & 4 \\ 1 & 4 & 0 \\ 1 & 4 & 0 \end{bmatrix}$ This product cycles the columns of **A**: column 1 moves to 2, column 2 moves to column 3, and column 3 moves to 1.

c. $AR = \begin{bmatrix} -1 & 9 & -4 \\ 3 & 12 & -2 \\ 4 & 0 & 2 \\ 4 & 0 & 2 \end{bmatrix}$ This multiplies the first column of **A** by 1, the second column by 3, and the third column by 2.

9. The price matrix **P** is

$$P = \begin{bmatrix} 0.30 & 0.25 & 0.30 \\ 0.20 & 0.25 & 0.25 \\ 0.35 & 0.40 & 0.30 \\ 0.50 & 0.60 & 0.45 \end{bmatrix}$$

with the different fruits in the rows and the different stores in the columns; the prices are in dollars. The purchase matrix is

$$A = \begin{bmatrix} 7 & 2 & 4 & 5 \\ 2 & 5 & 1 & 8 \end{bmatrix}$$

with the different fruits in the columns, Amy in row 1 and Bill in row 2. The matrix

$$AP = \begin{bmatrix} 6.40 & 6.85 & 6.05 \\ 5.95 & 6.95 & 5.75 \end{bmatrix}$$

gives the amount spent by each person at each store, in dollars.

11. The matrix **A** gives prices in dollars, with stores in the columns and types of fruit in the rows:

$$A = \begin{bmatrix} 0.15 & 0.20 \\ 0.25 & 0.15 \\ 0.20 & 0.25 \end{bmatrix}$$

The matrix **B** gives demand, with types of people in the rows and types of fruit in the columns:

$$B = \begin{bmatrix} 6 & 12 & 4 \\ 6 & 8 & 5 \end{bmatrix}$$

The matrix **C** gives residence by town, with towns in the rows and types of people in the columns:

$$C = \begin{bmatrix} 2000 & 800 \\ 1500 & 1200 \end{bmatrix}$$

a. To find how much each type of person spends on fruit at each store, we need a product that has people and stores on the "outside" and fruit as the common "inside" quantity, that is, the rows of the first matrix and the columns of the second matrix in the product. This will be **BA**:

$$BA = \begin{bmatrix} 4.70 & 4.00 \\ 3.90 & 3.65 \end{bmatrix}$$

The amounts are in dollars, with people in the rows and stores in the columns.

b. To find the amount of each type of fruit purchased in each town, we need a product relating fruit and towns, in which people will be the shared "inside" quantity. So we want (towns by people) times (people by fruit), giving (towns by fruit). This product is **CB**:

$$CB = \begin{bmatrix} 16,800 & 30,400 & 12,000 \\ 16,200 & 27,600 & 12,000 \end{bmatrix}$$

c. Assuming that each town is populated entirely by professors and engineers, we can find out how much was spend by each town at each store by computing
(towns by people) times (people by fruit) times (fruit by stores), giving (towns by stores). This is the product **CBA**:

$$CBA = \begin{bmatrix} 12,520 & 10,920 \\ 11,730 & 10,380 \end{bmatrix}$$

The entries are amounts in dollars.

13. The coefficient matrix for the goat-sheep-bear system is

$$A = \begin{bmatrix} 1 & 1 & 1 \\ 1 & 2 & -1 \\ 2 & -1 & 1 \end{bmatrix}.$$

a. $x_n = A^n x_0$

b. $A^2 = \begin{bmatrix} 4 & 2 & 1 \\ 1 & 6 & -2 \\ 3 & -1 & 4 \end{bmatrix}$, $A^3 = \begin{bmatrix} 8 & 7 & 3 \\ 3 & 15 & -7 \\ 10 & -3 & 8 \end{bmatrix}$

c. If $x_0 = \begin{bmatrix} 1 \\ 1 \\ 2 \end{bmatrix}$, $A^2 x_0 = \begin{bmatrix} 8 \\ 3 \\ 10 \end{bmatrix}$ and $A^3 x_0 = \begin{bmatrix} 21 \\ 4 \\ 23 \end{bmatrix}$.

d. The first year with a negative population is the fifth year, with

$$\mathbf{A}^5 \mathbf{x}_0 = \begin{bmatrix} 115 \\ -1 \\ 151 \end{bmatrix}.$$

Thus the model indicates that the sheep die out some time during the fourth year.

15. The transition matrix is

$$
\begin{array}{c}
\text{today} \\
\begin{array}{cc} \text{s} & \text{c} \end{array} \\
\mathbf{A} = \begin{bmatrix} \frac{3}{4} & \frac{1}{4} \\ \frac{1}{4} & \frac{3}{4} \end{bmatrix} \begin{array}{l} \text{sunny} \\ \text{cloudy} \end{array} \; \text{tomorrow}
\end{array}
$$

a. $\mathbf{A}^2 = \begin{bmatrix} \frac{5}{8} & \frac{3}{8} \\ \frac{3}{8} & \frac{5}{8} \end{bmatrix}$

The entry in position $(1, 2)$, which is $3/8$, represents the probability that if it is cloudy today it will be sunny the day after tomorrow.

b. $\mathbf{A}^3 = \begin{bmatrix} \frac{9}{16} & \frac{7}{16} \\ \frac{7}{16} & \frac{9}{16} \end{bmatrix}$

If it is cloudy today, the probability that it will be sunny in two days is $7/16$.

c. $\mathbf{A} = \begin{bmatrix} 0.504 & 0.496 \\ 0.496 & 0.504 \end{bmatrix}$

If it is cloudy today, the probability that it will be sunny a week from today is 0.496.

17. Using column vectors, we let

$$\mathbf{x} = \begin{bmatrix} 1 \\ x \\ x^2 \\ x^3 \end{bmatrix}, \quad \mathbf{p} = \begin{bmatrix} 2 \\ -7 \\ 4 \\ 1 \end{bmatrix}, \quad \mathbf{q} = \begin{bmatrix} 17 \\ 8 \\ -5 \\ 4 \end{bmatrix}.$$

Then $P(x) = \mathbf{p} \cdot \mathbf{x}$ and $Q(x) = \mathbf{q} \cdot \mathbf{x}$.

a. $P(x) + Q(x) = \mathbf{p} \cdot \mathbf{x} + \mathbf{q} \cdot \mathbf{x} = (\mathbf{p} + \mathbf{q}) \cdot \mathbf{x}$

b. $P(x) - Q(x) = \mathbf{p} \cdot \mathbf{x} - \mathbf{q} \cdot \mathbf{x} = (\mathbf{p} - \mathbf{q}) \cdot \mathbf{x}$

19. $\mathbf{R} = \begin{bmatrix} \cos\theta & -\sin\theta \\ \sin\theta & \cos\theta \end{bmatrix}$ Using double-angle identities we find:

$$\mathbf{R}^2 = \begin{bmatrix} \cos^2\theta - \sin^2\theta & -2\cos\theta\sin\theta \\ 2\cos\theta\sin\theta & \cos^2\theta - \sin^2\theta \end{bmatrix} = \begin{bmatrix} \cos 2\theta & -\sin 2\theta \\ \sin 2\theta & \cos 2\theta \end{bmatrix}$$

$$\mathbf{RR}^2 = \begin{bmatrix} \cos\theta\cos 2\theta - \sin\theta\sin 2\theta & -\cos\theta\sin 2\theta - \sin\theta\cos 2\theta \\ \sin\theta\cos 2\theta + \cos\theta\sin 2\theta & \cos\theta\cos 2\theta - \sin\theta\sin 2\theta \end{bmatrix}$$

Using formulas for the sine and cosine of the sum of two angles we can write this as

$$\begin{bmatrix} \cos(\theta + 2\theta) & -\sin(\theta + 2\theta) \\ \sin(\theta + 2\theta) & \cos(\theta + 2\theta) \end{bmatrix} = \begin{bmatrix} \cos 3\theta & -\sin 3\theta \\ \sin 3\theta & \cos 3\theta \end{bmatrix}.$$

21. Applying the nth power of R to any nonzero vector rotates that vector n degrees counterclockwise.

Section 10.5 Gaussian Elimination

1. We show the augmented matrix after each step.

a.
$$\begin{bmatrix} 1 & 1 & | & 8 \\ 2 & -3 & | & 4 \end{bmatrix}$$

Add -2 times the first equation to the second:
$$\begin{bmatrix} 1 & 1 & | & 8 \\ 0 & -5 & | & -12 \end{bmatrix}$$

Use back substitution:

$$-5y = 12 \qquad \text{so} \quad y = \frac{12}{5}$$

$$x + 1\left(\frac{12}{5}\right) = 8 \quad \text{so} \quad x = \frac{28}{5}$$

b.
$$\begin{bmatrix} 1 & -4 & | & 3 \\ -1 & 2 & | & 9 \end{bmatrix}$$

Add the first equation to the second:
$$\begin{bmatrix} 1 & -4 & | & 3 \\ 0 & -2 & | & 12 \end{bmatrix}$$

Use back substitution:
$$-2y = 12 \qquad \text{so} \quad y = -6$$
$$x - 4(-6) = 3 \quad \text{so} \quad x = -21$$

c.
$$\begin{bmatrix} 2 & -3 & | & 4 \\ -4 & 1 & | & 1 \end{bmatrix}$$

Add twice the first equation to the second:
$$\begin{bmatrix} 2 & -3 & | & 4 \\ 0 & -5 & | & 9 \end{bmatrix}$$

Use back substitution:

$$-5y = 9 \qquad\qquad \text{so} \quad y = -\frac{9}{5}$$

$$2x - 3\left(-\frac{9}{5}\right) = 4 \quad \text{so} \quad x = -\frac{7}{10}$$

d.
$$\begin{bmatrix} 3 & -2 & | & 3 \\ -2 & 5 & | & 2 \end{bmatrix}$$

Multiply the first equation by $\frac{1}{3}$:

$$\begin{bmatrix} 1 & -\frac{2}{3} & | & 1 \\ -2 & 5 & | & 2 \end{bmatrix}$$

Add twice the first equation to the second:

$$\begin{bmatrix} 1 & -\frac{2}{3} & | & 1 \\ 0 & \frac{11}{3} & | & 4 \end{bmatrix}$$

Use back substitution:

$$\frac{11}{3}y = 4 \qquad \text{so} \qquad y = \frac{12}{11}$$

$$x - \frac{2}{3}\left(\frac{12}{11}\right) = 1 \quad \text{so} \qquad x = \frac{19}{11}$$

e. $\begin{bmatrix} 3 & -2 & | & 3 \\ -4 & 3 & | & -2 \end{bmatrix}$

Multiply the first equation by $\frac{1}{3}$:

$$\begin{bmatrix} 1 & -\frac{2}{3} & | & 1 \\ -4 & 3 & | & -2 \end{bmatrix}$$

Add 4 times the first equation to the second:

$$\begin{bmatrix} 1 & -\frac{2}{3} & | & 1 \\ 0 & \frac{1}{3} & | & 2 \end{bmatrix}$$

Use back substitution:

$$\frac{1}{3}y = 2 \qquad \text{so} \quad y = 6$$

$$x - \frac{2}{3}(6) = 1 \quad \text{so} \quad x = 5$$

3. We show the augmented matrix after each step. When we refer to a row by number, we are referring to the most recent version of the system as shown in the immediately preceding augmented matrix. We omit the back substitution steps.

a. $\begin{bmatrix} 1 & 2 & 1 & | & 3 \\ -1 & 2 & 3 & | & 1 \\ 2 & -1 & 2 & | & 2 \end{bmatrix}$

Add row 1 to row 2:

$$\begin{bmatrix} 1 & 2 & 1 & | & 3 \\ 0 & 4 & 4 & | & 4 \\ 2 & -1 & 2 & | & 2 \end{bmatrix}$$

Add -2 times row 1 to row 3:

$$\begin{bmatrix} 1 & 2 & 1 & | & 3 \\ 0 & 4 & 4 & | & 4 \\ 0 & -5 & 0 & | & -4 \end{bmatrix}$$

Multiply row 2 by $\frac{1}{4}$:

$$\begin{bmatrix} 1 & 2 & 1 & | & 3 \\ 0 & 1 & 1 & | & 1 \\ 0 & -5 & 0 & | & -4 \end{bmatrix}$$

Add 5 times row 2 to row 3:

$$\begin{bmatrix} 1 & 2 & 1 & 3 \\ 0 & 1 & 1 & 1 \\ 0 & 0 & 5 & 1 \end{bmatrix}$$

The solution is

$$\begin{bmatrix} \frac{6}{5} \\ \frac{4}{5} \\ \frac{1}{5} \end{bmatrix};$$ that is, the solution is $x_1 = \frac{6}{5}$, $x_2 = \frac{4}{5}$, $x_3 = \frac{1}{5}$.

b. $\begin{bmatrix} 2 & -1 & 1 & 2 \\ -1 & -2 & 2 & -3 \\ 3 & 2 & -1 & 0 \end{bmatrix}$

Switch rows 1 and 2:

$$\begin{bmatrix} -1 & -2 & 2 & -3 \\ 2 & -1 & 1 & 2 \\ 3 & 2 & -1 & 0 \end{bmatrix}$$

Add 2 times row 1 to row 2:

$$\begin{bmatrix} -1 & -2 & 2 & -3 \\ 0 & -5 & 5 & -4 \\ 3 & 2 & -1 & 0 \end{bmatrix}$$

Add 3 times row 1 to row 3:

$$\begin{bmatrix} -1 & -2 & 2 & -3 \\ 0 & -5 & 5 & -4 \\ 0 & -4 & 5 & -9 \end{bmatrix}$$

Add $-\frac{4}{5}$ times row 2 to row 3:

$$\begin{bmatrix} -1 & -2 & 2 & -3 \\ 0 & -5 & 5 & -4 \\ 0 & 0 & 1 & -\frac{29}{5} \end{bmatrix}$$

The solution is

$$\begin{bmatrix} \frac{7}{5} \\ -5 \\ -\frac{29}{5} \end{bmatrix};$$ that is, the solution is $x_1 = \frac{7}{5}$, $x_2 = -5$, $x_3 = -\frac{29}{5}$.

c. $\begin{bmatrix} -1 & -3 & 2 & -1 \\ 5 & 4 & 6 & 12 \\ 2 & 1 & 3 & 4 \end{bmatrix}$

Add 5 times row 1 to row 2:

$$\begin{bmatrix} -1 & -3 & 2 & | & -1 \\ 0 & -11 & 16 & | & 7 \\ 2 & 1 & 3 & | & 4 \end{bmatrix}$$

Add 2 times row 1 to row 3:

$$\begin{bmatrix} -1 & -3 & 2 & | & -1 \\ 0 & -11 & 16 & | & 7 \\ 0 & -5 & 7 & | & 2 \end{bmatrix}$$

Add $-\dfrac{5}{11}$ times row 2 to row 3:

$$\begin{bmatrix} -1 & -3 & 2 & | & -1 \\ 0 & -11 & 16 & | & 7 \\ 0 & 0 & -\dfrac{3}{11} & | & -\dfrac{13}{11} \end{bmatrix}$$

The solution is

$$\begin{bmatrix} -\dfrac{22}{3} \\ \dfrac{17}{3} \\ \dfrac{13}{3} \end{bmatrix} ; \quad \text{that is, the solution is } x_1 = -\dfrac{22}{3}, \ x_2 = \dfrac{17}{3}, \ x_3 = \dfrac{13}{3}.$$

d. $\begin{bmatrix} 2 & 4 & -2 & | & 4 \\ -2 & 2 & -3 & | & -4 \\ 1 & -1 & -2 & | & -1 \end{bmatrix}$

Switch rows 1 and 3:

$$\begin{bmatrix} 1 & -1 & -2 & | & -1 \\ -2 & 2 & -3 & | & -4 \\ 2 & 4 & -2 & | & 4 \end{bmatrix}$$

Add 2 times row 1 to row 2:

$$\begin{bmatrix} 1 & -1 & -2 & | & -1 \\ 0 & 0 & -7 & | & -6 \\ 2 & 4 & -2 & | & 4 \end{bmatrix}$$

Add -2 times row 1 to row 3:

$$\begin{bmatrix} 1 & -1 & -2 & | & -1 \\ 0 & 0 & -7 & | & -6 \\ 0 & 6 & 2 & | & 6 \end{bmatrix}$$

Switch rows 2 and 3:

$$\begin{bmatrix} 1 & -1 & -2 & | & -1 \\ 0 & 6 & 2 & | & 6 \\ 0 & 0 & -7 & | & -6 \end{bmatrix}$$

The solution is

$$\begin{bmatrix} \frac{10}{7} \\ \frac{5}{7} \\ \frac{6}{7} \end{bmatrix}; \quad \text{that is, the solution is } x_1 = \frac{10}{7}, \ x_2 = \frac{5}{7}, \ x_3 = \frac{6}{7}.$$

e. $\begin{bmatrix} 1 & 2 & 2 & | & 3 \\ 1 & 2 & 3 & | & 11 \\ 2 & 1 & 2 & | & 5 \end{bmatrix}$

Add −1 times row 1 to row 2:

$\begin{bmatrix} 1 & 2 & 2 & | & 3 \\ 0 & 0 & 1 & | & 8 \\ 2 & 1 & 2 & | & 5 \end{bmatrix}$

Add −2 times row 1 to row 3:

$\begin{bmatrix} 1 & 2 & 2 & | & 3 \\ 0 & 0 & 1 & | & 8 \\ 0 & -3 & -2 & | & -1 \end{bmatrix}$

Switch rows 2 and 3:

$\begin{bmatrix} 1 & 2 & 2 & | & 3 \\ 0 & -3 & -2 & | & -1 \\ 0 & 0 & 1 & | & 8 \end{bmatrix}$

The solution is

$$\begin{bmatrix} -3 \\ -5 \\ 8 \end{bmatrix}; \quad \text{that is the solution is } x_1 = -3, \ x_2 = -5, \ x_3 = 8.$$

f. $\begin{bmatrix} 1 & -1 & -2 & | & 2 \\ 3 & 2 & 4 & | & 2 \\ 3 & 1 & -2 & | & -3 \end{bmatrix}$

Add −3 times row 1 to row 2:

$\begin{bmatrix} 1 & -1 & -2 & | & 2 \\ 0 & 5 & 10 & | & -4 \\ 3 & 1 & -2 & | & -3 \end{bmatrix}$

Add −3 times row 1 to row 3:

$\begin{bmatrix} 1 & -1 & -2 & | & 2 \\ 0 & 5 & 10 & | & -4 \\ 0 & 4 & 4 & | & -9 \end{bmatrix}$

Add $-\dfrac{4}{5}$ times row 2 to row 3:

$$\begin{bmatrix} 1 & -1 & -2 & 2 \\ 0 & 5 & 10 & -4 \\ 0 & 0 & -4 & -\dfrac{29}{5} \end{bmatrix}$$

The solution is

$$\begin{bmatrix} \dfrac{6}{5} \\ -\dfrac{37}{10} \\ \dfrac{29}{20} \end{bmatrix};\quad \text{that is, the solution is } x_1 = \dfrac{6}{5},\ x_2 = -\dfrac{37}{10},\ x_3 = \dfrac{29}{20}.$$

5. a. The second equation is a multiple of the first, so the system has multiple solutions.
   b. Multiplying the first equation by −3 produces identical left sides but different right sides, so the system is inconsistent.
   c. The second equation is a multiple of the first, so the system has multiple solutions.
   d. Gaussian elimination produces the augmented matrix $\begin{bmatrix} 1 & 2 & 3 & 10 \\ 0 & -5 & -2 & 0 \\ 0 & 0 & 0 & 0 \end{bmatrix}$. Since the last row is all 0's,

   the system has multiple solutions.
   e. Gaussian elimination produces the augmented matrix $\begin{bmatrix} 1 & 5 & 4 & 10 \\ 0 & -2 & 2 & -1 \\ 0 & 0 & -9 & -2 \end{bmatrix}$. This system has a unique

   solution.
   f. We can see by inspection that there is at least one solution, with all the variables equal to 0. Gaussian

   elimination produces the augmented matrix $\begin{bmatrix} -1 & 2 & 1 & 0 \\ 0 & 5 & -1 & 0 \\ 0 & 0 & 3.6 & 0 \end{bmatrix}$, so this is the only solution.

7. The system is $\mathbf{Ax} = \mathbf{b}$ with $\mathbf{A} = \begin{bmatrix} 6 & 4 & 2 \\ 4 & 8 & 4 \\ 3 & 2 & 8 \end{bmatrix}$ and $\mathbf{b} = \begin{bmatrix} 400 \\ 800 \\ 500 \end{bmatrix}$. The corresponding augmented matrix is

$\begin{bmatrix} 6 & 4 & 2 & 400 \\ 4 & 8 & 4 & 800 \\ 3 & 2 & 8 & 500 \end{bmatrix}$. Gaussian elimination leads to the augmented matrix $\begin{bmatrix} 1 & 2 & 1 & 200 \\ 0 & -8 & -4 & -800 \\ 0 & 0 & 7 & 300 \end{bmatrix}$ which

leads to a solution vector of $\begin{bmatrix} 0 \\ 78.57 \\ 42.86 \end{bmatrix}$, where the three entries give production levels for the three factories.

9.   The system is $\mathbf{Ax} = \mathbf{b}$ with $\mathbf{A} = \begin{bmatrix} 10 & 50 & 200 \\ 1 & 3 & 0.2 \\ 30 & 10 & 0 \end{bmatrix}$ and $\mathbf{b} = \begin{bmatrix} 600 \\ 20 \\ 200 \end{bmatrix}$. The corresponding augmented matrix

is $\begin{bmatrix} 10 & 50 & 200 & | & 600 \\ 1 & 3 & 0.2 & | & 20 \\ 30 & 10 & 0 & | & 200 \end{bmatrix}$. Gaussian elimination leads to the augmented matrix

$\begin{bmatrix} 1 & 3 & 0.2 & | & 20 \\ 0 & 20 & 198 & | & 400 \\ 0 & 0 & 786 & | & 1200 \end{bmatrix}$ which produces a solution vector $\begin{bmatrix} 5.04 \\ 4.89 \\ 1.53 \end{bmatrix}$, where the entries indicate the number

of units of gelatin, fish sticks, and mystery meat.

11.   To find the equilibrium states we need to solve $\mathbf{Ax} = \mathbf{x}$ or $\mathbf{Ax} - \mathbf{x} = \mathbf{0}$. The matrix for the left side of this

system is $\mathbf{A} - \begin{bmatrix} 1 & 0 \\ 0 & 1 \end{bmatrix}$ and the vector on the right side is $\begin{bmatrix} 0 \\ 0 \end{bmatrix}$.

a. The system has augmented matrix $\mathbf{A} = \begin{bmatrix} \frac{2}{3}-1 & \frac{1}{3} & | & 0 \\ \frac{1}{3} & \frac{2}{3}-1 & | & 0 \end{bmatrix}$ which has solution $x = y$. If we assume that

the state values add to 1, the equilibrium is thus $\begin{bmatrix} \frac{1}{2} \\ \frac{1}{2} \end{bmatrix}$.

b. The system has augmented matrix $\mathbf{A} = \begin{bmatrix} \frac{3}{4}-1 & \frac{1}{4} & | & 0 \\ \frac{1}{4} & \frac{3}{4}-1 & | & 0 \end{bmatrix}$ which has solution $x = y$. If we assume that

the state values add to 1, the equilibrium is $\begin{bmatrix} \frac{1}{2} \\ \frac{1}{2} \end{bmatrix}$.

c. The system has augmented matrix $\mathbf{A} = \begin{bmatrix} 1-1 & \frac{1}{2} & | & 0 \\ 0 & \frac{1}{2}-1 & | & 0 \end{bmatrix}$ which has solution $y = 0$. If we assume that

the state values add to 1, the equilibrium solution is thus $\begin{bmatrix} 1 \\ 0 \end{bmatrix}$.

13.   The transition matrix for this chain is $\begin{bmatrix} \frac{2}{3} & \frac{1}{2} \\ \frac{1}{3} & \frac{1}{2} \end{bmatrix}$ so to find the equilibrium, we solve the system represented

by $\begin{bmatrix} \frac{2}{3}-1 & \frac{1}{2} & | & 0 \\ \frac{1}{3} & \frac{1}{2}-1 & | & 0 \end{bmatrix}$. The solution is $x = \frac{3}{2} y$, which together with $x + y = 1$ gives the equilibrium vector

$\begin{bmatrix} \frac{3}{5} \\ \frac{2}{5} \end{bmatrix}$.

15.   The transition matrix for this chain is $\begin{bmatrix} \frac{1}{2} & \frac{1}{4} & 0 & 0 \\ \frac{1}{2} & \frac{1}{2} & \frac{1}{4} & 0 \\ 0 & \frac{1}{4} & \frac{1}{2} & 0 \\ 0 & 0 & \frac{1}{4} & 1 \end{bmatrix}$.   The augmented matrix for the system we need to

solve is then $\left[\begin{array}{cccc|c} \frac{1}{2}-1 & \frac{1}{4} & 0 & 0 & 0 \\ \frac{1}{2} & \frac{1}{2}-1 & \frac{1}{4} & 0 & 0 \\ 0 & \frac{1}{4} & \frac{1}{2}-1 & 0 & 0 \\ 0 & 0 & \frac{1}{4} & 1-1 & 0 \end{array}\right]$, which yields the solution $\begin{bmatrix} 0 \\ 0 \\ 0 \\ p_4 \end{bmatrix}$. This implies an equilibrium

state of $\begin{bmatrix} 0 \\ 0 \\ 0 \\ 1 \end{bmatrix}$, which corresponds to complete mastery.

17.   The three points are $(2,-1)$, $(4,3)$, and $(-1,8)$, so the system we need to solve for the coefficients a, b, and

c in $y = ax^2 + bx + c$ is:

$-1 = a(2^2) + b(2) + c$

$3 = a(4^2) + b(4) + c$

$8 = a(-1)^2 + b(-1) + c$

The equation is $y = x^2 - 4x + 3$.

19.   We use the equation $x^2 + y^2 + Cx + Dy + E = 0$ and solve for the coefficients C, D, and E.

   a.   The points are $(1,0)$, $(0,1)$, and $(1,1)$, so the system for the coefficients is:

$\left. \begin{array}{l} 1^2 + 0^2 + C(1) + D(0) + E = 0 \\ 0^2 + 1^2 + C(0) + D(1) + E = 0 \\ 1^2 + 1^2 + C(1) + D(1) + E = 0 \end{array} \right\} \Rightarrow C = -1,\ D = -1,\ E = 0$

The equation of the circle is $x^2 + y^2 - x - y = 0$.

   b.   The points are $(2,1)$, $(2,-3)$, and $(0,1)$, so the system for the coefficients is:

$\left. \begin{array}{l} 2^2 + 1^2 + C(2) + D(1) + E = 0 \\ 2^2 + (-3)^2 + C(2) + D(-3) + E = 0 \\ 0^2 + 1^2 + C(0) + D(1) + E = 0 \end{array} \right\} \Rightarrow C = -2,\ D = 2,\ E = -3$

The equation of the circle is $x^2 + y^2 - 2x + 2y - 3 = 0$.

c. The points are $(3, 1)$, $(2, 5)$, and $(-3, 6)$, so the system for the coefficients is:

$$\left. \begin{array}{l} 3^2 + 1^2 + C(3) + D(1) + E = 0 \\ 2^2 + 5^2 + C(2) + D(5) + E = 0 \\ (-3)^2 + 6^2 + C(-3) + D(6) + E = 0 \end{array} \right\} \Rightarrow C = \frac{45}{19},\ D = -\frac{79}{19},\ E = -\frac{246}{19}$$

The equation of the circle is $x^2 + y^2 + \frac{45}{19}x - \frac{79}{19}y - \frac{246}{19} = 0$.

d. The points are $(-1, -1)$, $(1, 3)$, and $(2, -1)$, so the system for the coefficients is:

$$\left. \begin{array}{l} (-1)^2 + (-1)^2 + C(-1) + D(-1) + E = 0 \\ 1^2 + 3^2 + C(1) + D(3) + E = 0 \\ 2^2 + (-1)^2 + C(2) + D(-1) + E = 0 \end{array} \right\} \Rightarrow C = -1,\ D = -\frac{3}{2},\ E = -\frac{9}{2}$$

The equation of the circle is $x^2 + y^2 - x - \frac{3}{2}y - \frac{9}{2} = 0$.

e. The points are $(0, 0)$, $(4, 1)$, and $(-4, 1)$, so the system for the coefficients is:

$$\left. \begin{array}{l} 0^2 + 0^2 + C(0) + D(0) + E = 0 \\ 4^2 + 1^2 + C(4) + D(1) + E = 0 \\ (-4)^2 + 1^2 + C(-4) + D(1) + E = 0 \end{array} \right\} \Rightarrow C = 0,\ D = -17,\ E = 0$$

The equation of the circle is $x^2 + y^2 - 17y = 0$.

21. If $A^{-1} = \begin{bmatrix} a & b \\ c & d \end{bmatrix}$, then for the matrix $A = \begin{bmatrix} 1 & 0 \\ 4 & 2 \end{bmatrix}$ we have

$$\begin{bmatrix} 1 & 0 \\ 4 & 2 \end{bmatrix} \begin{bmatrix} a & b \\ c & d \end{bmatrix} = \begin{bmatrix} a & b \\ 4a+2c & 4b+2d \end{bmatrix} = \begin{bmatrix} 1 & 0 \\ 0 & 1 \end{bmatrix},$$

which gives four equations for the four unknown entries that make up the inverse of A:

$$a = 1$$
$$b = 0$$
$$4a + 2c = 0$$
$$4b + 2d = 1$$

This implies that $c = -2$ and $d = \frac{1}{2}$. Thus $A^{-1} = \begin{bmatrix} 1 & 0 \\ -2 & 0.5 \end{bmatrix}$. This agrees with the calculator answer.

23. a. The calculator gives $y = ax + b$ with $a = 10.8$ and $b = 1.5333$.

    b. The completed table looks like this:

x	y	x^2	$x \cdot y$
1	11	1	11
2	25	4	50
3	33	9	99
4	45	16	180
5	57	25	285
6	65	36	390
Totals: 21	236	91	1015

    c. With the number of points n equal to 6, the system of linear equations for the coefficients a and b is:

$$21a + 6b = 236$$
$$91a + 21b = 1015$$

    d. The solution to this system is $a = \dfrac{54}{5}$ and $b = \dfrac{23}{15}$. These are the same coefficients we obtained in part (a).

    e. With large data values, the sums calculated in part (b) might get very large, and the equations in part (c) might involve numbers of widely varying sizes, which could lead to round off errors in the arithmetic.

Chapter 10 Review Problems

1. **a.** $A = \begin{bmatrix} 6 & 6 \end{bmatrix}$, $B = \begin{bmatrix} 2 & 3 \end{bmatrix}$

        $B - A = \begin{bmatrix} 2-6 & 3-6 \end{bmatrix} = \begin{bmatrix} -4 & -3 \end{bmatrix}$

        $\| B - A \| = \sqrt{(-4)^2 + (-3)^2} = 5$

    b. $A = \begin{bmatrix} -1 & 1 \end{bmatrix}$. $B = \begin{bmatrix} 4 & -1 \end{bmatrix}$

        $B - A = \begin{bmatrix} 4-(-1) & 1-(-1) \end{bmatrix} = \begin{bmatrix} 5 & 2 \end{bmatrix}$

        $\| B - A \| = \sqrt{5^2 + 2^2} = \sqrt{29}$

    c. $A = \begin{bmatrix} 1 & 2 & -3 \end{bmatrix}$, $B = \begin{bmatrix} 3 & 3 & -5 \end{bmatrix}$

        $B - A = \begin{bmatrix} 3-1 & 3-2 & -5-(-3) \end{bmatrix} = \begin{bmatrix} 2 & 1 & -2 \end{bmatrix}$

        $\| B - A \| = \sqrt{2^2 + 1^2 + (-2)^2} = 3$

2. a.

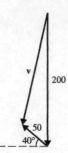

We choose a coordinate system with east along the positive x-axis and north along the positive y-axis.
The plane's heading is $\mathbf{h} = \begin{bmatrix} 0 & -200 \end{bmatrix}$ and the wind's velocity is $\mathbf{w} = \begin{bmatrix} -50\cos 40° & 50\sin 40° \end{bmatrix}$
$= \begin{bmatrix} -38.30 & 32.14 \end{bmatrix}$. The plane's actual velocity is $\mathbf{h} + \mathbf{w} = \begin{bmatrix} -38.30 & -167.86 \end{bmatrix}$. To find the direction
we find $\theta = \arctan\left(\dfrac{167.86}{38.30}\right) = 77.15°$, so the direction is 77.15° south of west. The plane's actual speed
is $\left\| \begin{bmatrix} -38.3 & -167.86 \end{bmatrix} \right\| = 172.17$ mph.

b. The new wind vector is $\mathbf{w} = \begin{bmatrix} 50\cos 40° & -50\sin 40° \end{bmatrix} = \begin{bmatrix} 38.30 & -32.14 \end{bmatrix}$. Thus the plane's actual

velocity vector is $\mathbf{h} + \mathbf{w} = \begin{bmatrix} 38.3 & -232.14 \end{bmatrix}$. Then $\theta = \arctan\left(\dfrac{-232.14}{38.30}\right) = -80.63°$. The direction is

80.63° south of east. The actual speed is $\left\| \begin{bmatrix} 38.3 & -232.14 \end{bmatrix} \right\| = 235.28$ mph.

3. If t, c, and s stand for the number of tables, chairs, and sofas produced, a system that determines production
levels based on available materials and labor is:
$$4t + c + 3s = 1500$$
$$3t + 2c + 5s = 2300$$
$$2c + 4s = 1800$$

4. a. The transition matrix is:

$$
\begin{array}{cccc}
 & \text{today} & & \\
U & E & D & \\
\begin{bmatrix} 0.6 & 0.4 & 0.2 \\ 0.2 & 0.2 & 0.2 \\ 0.2 & 0.4 & 0.6 \end{bmatrix} & \begin{array}{l} \text{up} \\ \text{even} \\ \text{down} \end{array} & \text{tomorrow}
\end{array}
$$

U up
E unchanged
D down

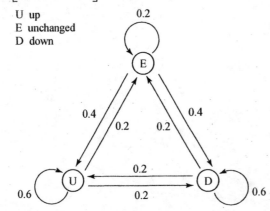

b. The distribution of probabilities for tomorrow is:

$$\begin{bmatrix} 0.6 & 0.4 & 0.2 \\ 0.2 & 0.2 & 0.2 \\ 0.2 & 0.4 & 0.6 \end{bmatrix}\begin{bmatrix} 0.3 \\ 0.1 \\ 0.6 \end{bmatrix} = \begin{bmatrix} 0.34 \\ 0.20 \\ 0.46 \end{bmatrix} \begin{matrix} u \\ e \\ d \end{matrix}$$

5. $\mathbf{a} = \begin{bmatrix} 1 \\ 2 \\ 3 \end{bmatrix}, \quad \mathbf{b} = \begin{bmatrix} -1 \\ 3 \\ -1 \end{bmatrix}, \quad \mathbf{c} = \begin{bmatrix} 2 \\ 5 \\ 8 \end{bmatrix}$

a. $\mathbf{a} \cdot \mathbf{b} = 1(-1) = 2(3) + 3(-1) = 2$

b. $\mathbf{b} \cdot \mathbf{c} = (-1)(2) + 3(5) + (-1)(8) = 5$

c. $\mathbf{a} \cdot (\mathbf{b} + \mathbf{c}) = 1(-1+2) + 2(3+5) + 3(-1+8) = 38$

d. $\mathbf{a} \cdot \mathbf{a} = 1(1) + 2(2) + 3(3) = 14$

6. **a**, **b**, and **c** are as in Problem 5.

$$A = \begin{bmatrix} 1 & 2 & 3 & 4 \\ 2 & 4 & 6 & 8 \\ 3 & 5 & 7 & 9 \end{bmatrix}, \quad B = \begin{bmatrix} 1 & 0 & -1 \\ 2 & -2 & 0 \\ 0 & 1 & 1 \end{bmatrix}, \quad C = \begin{bmatrix} 5 & 4 & 1 \\ 1 & 0 & 2 \\ 3 & 2 & 1 \\ 0 & 1 & 3 \end{bmatrix}$$

a. **aA** is not defined.
b. **bB** is not defined.
c. **cC** is not defined.
d. **Aa** is not defined.

e. $Bb = \begin{bmatrix} 1 & 0 & -1 \\ 2 & -2 & 0 \\ 0 & 1 & 1 \end{bmatrix}\begin{bmatrix} -1 \\ 3 \\ -1 \end{bmatrix} = \begin{bmatrix} 1(-1)+0(3)+(-1)(-1) \\ 2(-1)+(-2)(3)+0(-1) \\ 0(-1)+1(3)+1(-1) \end{bmatrix} = \begin{bmatrix} 0 \\ -8 \\ 2 \end{bmatrix}$

f. $Cc = \begin{bmatrix} 5 & 4 & 1 \\ 1 & 0 & 2 \\ 3 & 2 & 1 \\ 0 & 1 & 3 \end{bmatrix}\begin{bmatrix} 2 \\ 5 \\ 8 \end{bmatrix} = \begin{bmatrix} 5(2)+4(5)+1(8) \\ 1(2)+0(5)+2(8) \\ 3(2)+2(5)+1(8) \\ 0(2)+1(5)+3(8) \end{bmatrix} = \begin{bmatrix} 38 \\ 18 \\ 24 \\ 29 \end{bmatrix}$

7. The matrix **A** gives the number of each circuit needed in each type of computer. The rows are computers, the columns are circuits. The row vector **d** gives the demand for each type of computer and the column vector **p** gives the price of each of the different circuits.

a. The total cost is given by **dAp**.

b. $\mathbf{dA} = \begin{bmatrix} 210 & 110 & 140 & 110 \end{bmatrix}$, so **dAp** is

$$\begin{bmatrix} 210 & 110 & 140 & 110 \end{bmatrix}\begin{bmatrix} 2 \\ 5 \\ 1 \\ 10 \end{bmatrix} = 210(2) + 110(5) + 140(1) + 110(10) = 2210.$$

The total cost is $2210.

8. $A = \begin{bmatrix} 1 & 2 & 3 & 4 \\ 2 & 4 & 6 & 8 \\ 3 & 5 & 7 & 9 \end{bmatrix}$, $B = \begin{bmatrix} 1 & 0 & -1 \\ 2 & -2 & 0 \\ 0 & 1 & 1 \end{bmatrix}$, $C = \begin{bmatrix} 5 & 4 & 1 \\ 1 & 0 & 2 \\ 3 & 2 & 1 \\ 0 & 1 & 3 \end{bmatrix}$

a. **AB** is not defined.

b. $BA = \begin{bmatrix} 1 & 0 & -1 \\ 2 & -2 & 0 \\ 0 & 1 & 1 \end{bmatrix}\begin{bmatrix} 1 & 2 & 3 & 4 \\ 2 & 4 & 6 & 8 \\ 3 & 5 & 7 & 9 \end{bmatrix}$

$= \begin{bmatrix} 1(1)+0(2)+(-1)(3) & 1(2)+0(4)+(-1)(5) & 1(3)+0(6)+(-1)(7) & 1(4)+0(8)+(-1)(9) \\ 2(1)+(-2)(2)+0(3) & 2(2)+(-2)(4)+0(5) & 2(3)+(-2)(6)+0(7) & 2(4)+(-2)(8)+0(9) \\ 0(1)+1(2)+1(3) & 0(2)+1(4)+1(5) & 0(3)+1(6)+1(7) & 0(4)+1(8)+1(9) \end{bmatrix}$

$= \begin{bmatrix} -2 & -3 & -4 & -5 \\ -2 & -4 & -6 & -8 \\ 5 & 9 & 13 & 17 \end{bmatrix}$

c. $AC = \begin{bmatrix} 16 & 14 & 20 \\ 32 & 28 & 40 \\ 41 & 35 & 47 \end{bmatrix}$

d. $CA = \begin{bmatrix} 16 & 31 & 46 & 61 \\ 7 & 12 & 17 & 22 \\ 10 & 19 & 28 & 37 \\ 11 & 19 & 27 & 35 \end{bmatrix}$

e. $CB = \begin{bmatrix} 13 & -7 & -4 \\ 1 & 2 & 1 \\ 7 & -3 & -2 \\ 2 & 1 & 3 \end{bmatrix}$

9. **A** gives prices, with fruit in rows, stores in columns; **B** gives fruit quantities bought, with people in rows, fruit in columns; **C** gives population, with towns in rows and types of people in columns.

$A = \begin{bmatrix} 0.10 & 0.15 \\ 0.15 & 0.20 \\ 0.10 & 0.10 \end{bmatrix}$, $B = \begin{bmatrix} 5 & 10 & 3 \\ 4 & 5 & 5 \end{bmatrix}$, $C = \begin{bmatrix} 1000 & 500 \\ 2000 & 1000 \end{bmatrix}$

a. The matrix $BA = \begin{bmatrix} 2.30 & 3.05 \\ 1.65 & 2.10 \end{bmatrix}$ gives the cost in dollars of each type of person's purchases at each store, with people in rows and stores in columns.

b. The matrix $CB = \begin{bmatrix} 7000 & 12500 & 5500 \\ 14,000 & 25,000 & 11,000 \end{bmatrix}$ gives the quantity of fruit bought in each town, with towns in rows and types of fruit in columns.

10. a. $A = \begin{bmatrix} 0.6 & 0.4 & 0.2 \\ 0.2 & 0.2 & 0.2 \\ 0.2 & 0.4 & 0.6 \end{bmatrix}$, $A^2 = \begin{bmatrix} 0.48 & 0.40 & 0.32 \\ 0.20 & 0.20 & 0.20 \\ 0.32 & 0.40 & 0.48 \end{bmatrix}$, $A^3 = \begin{bmatrix} 0.432 & 0.400 & 0.368 \\ 0.200 & 0.200 & 0.200 \\ 0.368 & 0.400 & 0.432 \end{bmatrix}$

$A^5 = \begin{bmatrix} 0.40512 & 0.40000 & 0.39488 \\ 0.20000 & 0.20000 & 0.20000 \\ 0.39488 & 0.40000 & 0.40512 \end{bmatrix}$

  b. The columns appear to be approaching the vector $\begin{bmatrix} 0.4 \\ 0.2 \\ 0.4 \end{bmatrix}$.

11. a. The augmented matrix for the system is

$\begin{bmatrix} 2 & -3 & 2 & | & 0 \\ 1 & -1 & 1 & | & 7 \\ -1 & 5 & 4 & | & 4 \end{bmatrix}$

Switch row1 and row 2:

$\begin{bmatrix} 1 & -1 & 1 & | & 7 \\ 2 & -3 & 2 & | & 0 \\ -1 & 5 & 4 & | & 4 \end{bmatrix}$

Add -2 times row 1 to row 2:

$\begin{bmatrix} 1 & -1 & 1 & | & 7 \\ 0 & -1 & 0 & | & -14 \\ -1 & 5 & 4 & | & 4 \end{bmatrix}$

Add row 1 to row 3:

$\begin{bmatrix} 1 & -1 & 1 & | & 7 \\ 0 & -1 & 0 & | & -14 \\ 0 & 4 & 5 & | & 11 \end{bmatrix}$

Add 4 times row 2 to row 3:

$\begin{bmatrix} 1 & -1 & 1 & | & 7 \\ 0 & -1 & 0 & | & -14 \\ 0 & 0 & 5 & | & -45 \end{bmatrix}$

Using back substitution we now find the solution:

$5x_3 = -45$ so $x_3 = -9$

$(-1)x_2 = -14$ so $x_2 = 14$

$x_1 - 14 + (-9) = 7$ so $x_1 = 30$

  b. The augmented matrix for the system is:

$\begin{bmatrix} -1 & -1 & 1 & | & 2 \\ 2 & 2 & -4 & | & -4 \\ 1 & -2 & 3 & | & 5 \end{bmatrix}$

Add twice row 1 to row 2:

$\begin{bmatrix} -1 & -1 & 1 & | & 2 \\ 0 & 0 & -2 & | & 0 \\ 1 & -2 & 3 & | & 5 \end{bmatrix}$

Add row 1 to row 3:

$$\begin{bmatrix} -1 & -1 & 1 & | & 2 \\ 0 & 0 & -2 & | & 0 \\ 0 & -3 & 4 & | & 7 \end{bmatrix}$$

Switch row 2 and row 3:

$$\begin{bmatrix} -1 & -1 & 1 & | & 2 \\ 0 & -3 & 4 & | & 7 \\ 0 & 0 & -2 & | & 0 \end{bmatrix}$$

The last line shows that $x_3 = 0$. Then

$$-3x_2 = 7 \qquad \text{so} \quad x_2 = -\frac{7}{3}$$

$$-x_1 - \left(-\frac{7}{3}\right) = 2 \quad \text{so} \quad x_1 = \frac{1}{3}$$

12. The system from Problem 3 is:

$$4t + c + 3s = 1500$$
$$3t + 2c + 5s = 2300$$
$$2c + 4s = 1800$$

The solution vector is $\begin{bmatrix} 100 \\ 500 \\ 200 \end{bmatrix}$; that is 100 tables, 500 chairs, and 200 sofas need to be produced.

13. To find the stable distribution for the Markov chain in Problem 4, we need to solve the system with the following augmented matrix:

$$\begin{bmatrix} -0.4 & 0.4 & 0.2 & | & 0 \\ 0.2 & -0.8 & 0.2 & | & 0 \\ 0.2 & 0.4 & -0.4 & | & 0 \end{bmatrix}$$

The solution is $\begin{bmatrix} p_3 \\ \dfrac{p_3}{2} \\ p_3 \end{bmatrix}$ and since these probabilities add to 1 the stable distribution is $\begin{bmatrix} 0.4 \\ 0.2 \\ 0.4 \end{bmatrix}$.

14. Since the points are (1, 1), (2, 2,) and (3, 5) the system for the coefficients a, b, and c in the equation

$$\left. \begin{array}{l} 1 = a(1^2) + b(1) + c \\ 2 = a(2^2) + b(2) + c \\ 5 = a(3^2) + b(3) + c \end{array} \right\} \Rightarrow \quad a = 1, \ b = -2, \ c = 2$$

Solving this system we find the equation $y = x^2 - 2x + 2$.